Advanced Optical Communication Systems

Advanced Optical Communication Systems

Edited by **Vladimir Latinovic**

New Jersey

Published by Clanrye International,
55 Van Reypen Street,
Jersey City, NJ 07306, USA
www.clanryeinternational.com

Advanced Optical Communication Systems
Edited by Vladimir Latinovic

© 2015 Clanrye International

International Standard Book Number: 978-1-63240-023-9 (Hardback)

Contents

Permissions

List of Contributors

Preface

This book presents a descriptive account on various latest advances in the field of optical communication systems. These systems are extremely essential for all types of networks and telecommunications. They comprise of a transmitter which encodes the message into an optical signal, a channel which carries the signal to its destination, and a receiver which reproduces the message from the received optical signal. This book provides updated results on communication systems, with elucidations on their relevance, provided by veteran researchers in this field. It encompasses fundamental concepts of optical and wireless optical communication systems, optical multiplexing and demultiplexing for optical communication systems, optical amplifiers and networks, and network traffic engineering. Lately, wavelength conversion and other advanced signal processing functions have also been studied in detail for optical communications systems. The book emphasizes on wavelength conversion, demultiplexing in the time domain, switching and other optimized functions for optical communications systems. It is primarily aimed at assisting in advancement and research for a wide range of readers including design engineer teams in manufacturing industry, academia and telecommunications service operators/providers.

The information shared in this book is based on empirical researches made by veterans in this field of study. The elaborative information provided in this book will help the readers further their scope of knowledge leading to advancements in this field.

Finally, I would like to thank my fellow researchers who gave constructive feedback and my family members who supported me at every step of my research.

Editor

Part 1

Optical Communications Systems:
General Concepts

Wireless Optical Communications Through the Turbulent Atmosphere: A Review

Ricardo Barrios and Federico Dios
Department of Signal Theory and Communications, Technical University of Catalonia
Spain

1. Introduction

In the past decades a renewed interest has been seen around optical wireless communications, commonly known as free-space optics (FSO), because of the ever growing demand for high-data-rate data transmission as to a large extent current applications, such as the high-definition (HD) contents and cloud computing, require great amount of data to be transmitted, hence, demanding more transmission bandwidth. Nowadays, the last mile problem continuous to be the bottle neck in the global communication network. While the fiber-optic infrastructure, commonly called network backbone, is capable of coping with current demand, the end user accesses the network data stream through copper based connection and radio-frequency (RF) wireless services, that are inherently slower technologies. As the number of user increases, the radio-frequency spectrum is getting so crowded that there is virtually no room for new wireless services within the RF band, with the added inconvenient of limited bandwidth restriction when using a RF band and the license fees that have to be paid in order to use such a band. Regarding cooper-based technologies and the lower-speed connections, compared with the backbone, that are offered such as DSl (digital line subscriber), cable modems, or T1's (transmission system 1), they are alternatives that makes the service provider to incur in extra installation costs for deploying the wired network through the city.

When a fiber-optic link is neither practical nor feasible, under the above scenario, wireless optical communications (WOC) becomes a real alternative, since it allows to transfer data with high-bandwidth requirements with the additional advantages of wireless systems (Arimoto, 2010; Ciaramella et al., 2009; Sova et al., 2006). Moreover, a wireless optical communication system offers, when compared with RF technology, an intrinsic narrower beam; less power, mass and volume requirements, and the advantage of no regulatory policies for using optical frequencies and bandwidth.

On the other hand, satellite communication systems is a field where FSO is becoming more attractive thanks to the advantages mentioned above, and the additional fact that for *satellite-satellite* links there is no beam degradation due to the absence of atmosphere. Nevertheless, the pointing system complexity is increased as the order of the optical beam divergence is hundreds of µrad, whereas for an RF beam is in the order of tens to hundreds of mrad. The Semi-Conductor Inter Satellite Link EXperiment (SILEX) was the first European project to conduct a successful demo with the transmission of data through an optical link, between the SPOT-4 and Artemis satellites achieving 50Mbps of transfer rate (Fletcher et al., 1991). There has also been experiments for *ground-satellite* optical links such as the Ground/Orbiter Lasercomm Demonstrator (GOLD) (Jeganathan et al., 1997); and for *air-satellite* link with the Airbone Atmospheric Laser Link (LOLA, for its French initials), which used of the Artemis optical payload and an airborne optical transceiver flying at 9000m (Cazaubiel et al., 2006).

The major drawback for deploying wireless links based on FSO technology, where lasers are used as sources, is the perturbation of the optical wave as it propagates through the turbulent atmosphere. Moreover, fog, rain, snow, haze, and generally any floating particle can cause extinction of the signal-carrying laser beam intensity. In a worst case scenario the intensity attenuation can be strong enough to cause link outages, leading to a high bit error-rate that inevitably decrease the overall system performance and limits the maximum length for the optical link.

The turbulent atmosphere produces many effects, of which the most noticeable is the random fluctuations of the traveling wave irradiance, phenomenon known as scintillation. Additionally, there are other effects that perturb the traveling wavefront such as beam wander, that is a continuous random movement of the beam centroid over the receiving aperture; angle-of-arrival fluctuations, which are associated with the dancing of the focused spot on the photodetector surface; and beam spreading that is the spreading beyond the pure diffraction limit of the beam radius.

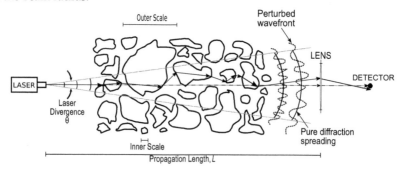

Fig. 1. Laser beam propagation through the turbulent atmosphere.

A compound of various perturbations suffered by an optical traveling wavefront is shown in Fig. 1. Here, the minor spatial scale distortions in the wavefronts provoke a random pattern, both in time and space, of self-interference of the beam at the receiver plane. As a consequence rapid variations of the received power appear, which is the most noticeable effect of the atmospheric turbulence over the optical link. The rays (solid arrows) leaving the laser source are deflected as they travel through the largest air pockets, whose size defines the turbulence outer scale, arriving off-axis instead of what is expected without turbulence, represented with the straight dashed arrow starting at laser and finishing at the receptor surface in Fig. 1. Additionally, the turbulent atmosphere induces a spreading of the beam that is the broadening of the beam size beyond of that expected due to pure diffraction, for the case of a laser beam. It is customary to refer as refractive effects to those caused by the outer scale size of turbulence, whereas, the inner scale sizes produces the diffractive effects. As the rays may also be interpreted as the wave vector for the traveling wavefront, the variations in the angle respect the optical axis at the receiver represent the concept of angle-of-arrival fluctuations. Furthermore, this bouncing of the optical wavefront as it propagates through the atmosphere is also responsible for the beam wander effect as the centroid of the laser beam is displaced randomly at the receiver plane.

This chapter is organized as follows: In Section 2 the most widely spread power spectrum models to characterize the turbulent atmosphere are addressed. Secondly, in Section 3, a short yet complete review of the propagation of optical electromagnetic waves in turbulent media is presented, followed by a brief introduction to the beam split-step method for the simulation of traveling optical beams in Section 4. Finally, WOC systems are addressed from a communication theory approach where general system characterization and performance evaluation are made in Section 5 and Section 6,respectively.

2. Atmospheric turbulence

All the models used to describe the effects of the atmosphere on a optical traveling wave are based on the study of turbulence, which involves fluctuations in the velocity field of a viscous fluid (Andrews & Philips, 2005). These variations in the fluid—the air for the atmosphere case—are firstly due to temperature differences between the surface of the Earth and the atmosphere, then, to the differences in temperature and pressure within the atmospheric layers themselves, thus, producing pockets of air, also known as eddies, that cause the atmospheric turbulence.

The different eddy sizes, called the inertial range, responsible for the transfer of kinetic energy within the fluid go from the outer scale L_0 to the inner scale l_0 of turbulence, where typical values of L_0 are between 10 and 20m, while l_0 is usually around 1-2mm. Such conditions comprise a continuum where wind energy is injected in the macroscale L_0, transfer through the inertial range and finally dissipated in the microscale l_0. This energy transfer causes unstable air masses, with temperature gradients, giving rise to local changes in the atmospheric refractive-index and thus creating what is called *optical turbulence* as an optical wave propagates. Treating the atmospheric turbulence as a consequence of the fluctuations in refractive-index instead of temperature is the natural way to address wave propagation for optical frequencies. Following this reasoning is a good approach to define a power spectral density for refractive-index fluctuations as a mean to express the atmospheric turbulence.

The variations of the atmospheric refractive-index n, which can be considered as locally homogeneous, can be mathematically expressed by

$$n(\vec{r}, t) = n_0 + n_1(\vec{r}, t), \tag{1}$$

where n_0 is the mean value of the index of refraction; $n_1(\vec{r}, t)$ is a random variable with zero mean, representing the changes caused by the atmospheric turbulence, and t indicates the temporal dependence. Nevertheless, under the *Taylor frozen turbulence hypothesis*, the turbulence is regarded as stationary as the optical wave propagates, hence, the time dependence is traditionally dropped in Eq. (1).

The statistical characterization of a locally homogeneous random field is usually done by its structure function, denoted by

$$D_n(\vec{r_1}, \vec{r_2}) = \langle [n(\vec{r_1}) - n(\vec{r_2})]^2 \rangle, \tag{2}$$

where there is no time dependence in the index of refraction.

2.1 Refractive-index structure parameter

The atmospheric turbulence can be defined by the strength of the fluctuations in the refractive-index, represented with the refractive-index structure parameter C_n^2 in units of $\text{m}^{-2/3}$—which has a direct relation with the structure function mentioned above. Along the optical propagation distance the value of C_n^2 have small variations for horizontal paths, while for slant and vertical paths these variations become significant. It's very common to assume a constant value for horizontal links, and to measure the path averaged value of C_n^2 from methods that rely on the atmospheric data *in situ* (Andreas, 1988; Doss-Hammel et al., 2004; Lawrence et al., 1970; Sadot & Kopeika, 1992), or others that extract the C_n^2 value from experimental scintillation data (Fried, 1967; Wang et al., 1978).

On the other hand, when a vertical path is considered, the behavior of C_n^2 is conditioned by temperature changes along the different layers within the Earth's atmosphere, hence, the refractive-index structure parameter becomes a function of the altitude above ground.

Model	Expression
SLC-Day	$C_n^2(h) = \begin{cases} 1.700 \cdot 10^{-14} & 0\text{m} < h < 19\text{m} \\ 4.008 \cdot 10^{-13} h^{-1.054} & 19\text{m} < h < 230\text{m} \\ 1.300 \cdot 10^{-15} & 230\text{m} < h < 850\text{m} \\ 6.352 \cdot 10^{-7} h^{-2.966} & 850\text{m} < h < 7000\text{m} \\ 6.209 \cdot 10^{-16} h^{-0.6229} & 7000\text{m} < h < 20000\text{m} \end{cases}$
Hufnagel-Valley Day	$C_n^2(h) = A e^{\frac{-h}{100}} + 5.94 \cdot 10^{-53} \left(\frac{v}{27}\right)^2 h^{10} e^{\frac{-h}{1000}} + 2.7 \cdot 10^{-16} e^{\frac{-h}{1500}}$
Hufnagel-Valley Night	$C_n^2(h) = 1.9 \cdot 10^{-15} e^{\frac{-h}{100}} + 8.16 \cdot 10^{-54} h^{10} e^{\frac{-h}{1000}} + 3.02 \cdot 10^{-17} e^{\frac{-h}{1500}}$
Greenwood	$C_n^2(h) = \left[2.2 \cdot 10^{-13}(h+10)^{-1.3} + 4.3 \cdot 10^{-17}\right] e^{-h/1500}$

Table 1. Refractive-index structure parameter models as a function of the altitude h above ground. For the HV model, $A = C_n^2(0)$ is the refractive-index structure parameter at ground level, and v is the root-mean-square wind speed.

Many authors have tried to predict the behavior of the refractive-index structure parameter, and various models have been proposed. However, it should be noted that most of these models are based on fittings from experiments conducted in specific places, which makes difficult their generalization. Table 1 presents a list of different C_n^2 models, namely, the Submarine Laser Communication (SLC) Day model and the Hufnagel-Valley, best suited for inland day-time conditions, the HV-Night for night-time conditions, and the Greenwood model adapted for astronomical tasks from mountaintop locations. A comparative of all four refractive-index structure parameter models is shown below in Fig. 2, where it is readily seen that day-time models predict higher values of C_n^2 than night-time models, as expected.

Sadot & Kopeika (1992) have developed an empirical model for estimating the refractive-index structure parameter from macroscale meteorological measurements *in situ*. The value of C_n^2 depends strongly on the hour of the day. It has a peak value at midday and local minima at sunrise and sunset. Provided that the time elapsed between the sunrise and sunset is different according to seasonal variations, the concept of *temporal hour* (t_h) has been introduced. The duration of a temporal hour is $1/12^\text{th}$ of the time between sunrise and

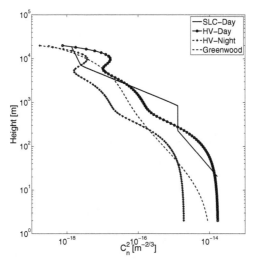

Fig. 2. Refractive-index structure parameter altitude profile of different models. For HV-day model $A = 1.7 \cdot 10^{-14} \mathrm{m}^{-2/3}$ and $v = 21 \mathrm{m/s}$.

sunset. In summer it is more than 60 min and in winter is smaller, therefore, it can be seen as a solar hour. The current t_h is obtained by subtracting the sunrise time from the local time, and dividing by the value of one t_h. Thus, in any day of the year $t_h = 00 : 00$ at sunrise, $t_h = 06 : 00$ at noon, and $t_h = 12 : 00$ at sunset. It should be noted that temporal hours are allowed to have negative time hours.

	Temporal hour interval		W_{t_h}		Temporal hour interval		W_{t_h}
	until	-4	0.11		5 to	6	1.00
	-4 to	-3	0.11		6 to	7	0.90
	-3 to	-2	0.07		7 to	8	0.80
	-2 to	-1	0.08		8 to	9	0.59
	-1 to	0	0.06		9 to	10	0.32
Sunrise →	0 to	1	0.05		10 to	11	0.22
	1 to	2	0.10	Sunset →	11 to	12	0.10
	2 to	3	0.51		12 to	13	0.08
	3 to	4	0.75		over	13	0.13
	4 to	5	0.95				

Table 2. Weight W parameter as a function of the corresponding temporal hour.

The expression obtained that describes C_n^2 is based on a polynomial regression model according to

$$C_n^2 = 3.8 \times 10^{-14} W_{t_h} + 2 \times 10^{-15} T - 2.8 \times 10^{-15} RH + 2.9 \times 10^{-17} RH^2 - 1.1 \times 10^{-19} RH^3$$
$$-2.5 \times 10^{-15} WS + 1.2 \times 10^{-15} WS^2 - 8.5 \times 10^{-17} WS^3 - 5.3 \times 10^{-13}, \tag{3}$$

where W_{t_h} denotes a temporal-hour weight (see Table 2), T is the temperature in Kelvins, RH is the relative humidity (%), and WS is the wind speed in $\mathrm{m\,s}^{-1}$.

An improved version of this model is also presented in Sadot & Kopeika (1992), with the introduction of the effects of solar radiation and aerosol loading in the atmosphere, as follows

$$C_n^2 = 5.9 \times 10^{-15}W_{t_h} + 1.6 \times 10^{-15}T - 3.7 \times 10^{-15}\text{RH} + 6.7 \times 10^{-17}\text{RH}^2$$
$$- 3.9 \times 10^{-19}\text{RH}^3 - 3.7 \times 10^{-15}\text{WS} + 1.3 \times 10^{-15}\text{WS}^2 - 8.2 \times 10^{-17}\text{WS}^3$$
$$+ 2.8 \times 10^{-14}\text{SF} - 1.8 \times 10^{-14}\text{TCSA} + 1.4 \times 10^{-14}\text{TCSA}^2 - 3.9 \times 10^{-13}, \qquad (4)$$

where SF is the solar flux in units of kW m^{-2}, and TCSA is the total cross-sectional area of the aerosol particles and its expression can be found in Yitzhaky et al. (1997)

$$TCSA = 9.96 \times 10^{-4}\text{RH} - 2.75 \times 10^{-5}\text{RH}^2 + 4.86 \times 10^{-7}\text{RH}^3 - 4.48 \times 10^{-9}\text{RH}^4$$
$$+ 1.66 \times 10^{-11}\text{RH}^5 - 6.26 \times 10^{-3}\ln \text{RH} - 1.37 \times 10^{-5}\text{SF}^4 + 7.30 \times 10^{-3}. \qquad (5)$$

A 24-hour data set of macroscale meteorological measurements taken at the Campus Nord in the Technical University of Catalonia in Barcelona, Spain, collected on the 14th of November of 2009 was used to generate the plot, shown in Fig. 3, of the estimated refractive-index structure parameter C_n^2 using Eq. (3) and Eq. (4) labeled as Model 1 and Model 2, respectively.

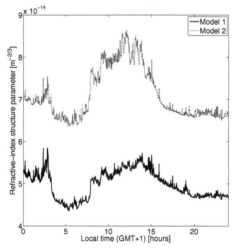

Fig. 3. Refractive-index structure parameter predicted from macroscale meteorological and aerosols data collected in an autumn day in Barcelona, Spain.

2.2 Atmospheric power spectrum models

The first studies on the atmospheric turbulence effects on propagating light waves were conducted by Tatarskii (1971) using the Rytov method and considering, as still does nowadays, the Kolmogorov turbulence spectrum (Kolmogorov, 1941) that suggest that the inertial range has a degree of statistical consistency, where points in the atmosphere separated certain scale size exhibit statistical homogeneity and isotropy. The use of these characteristics, along with additional simplifications and assumptions, were essential to develop tractable expressions for a fundamentally nonlinear phenomenon, as the atmospheric turbulence.

Kolmogorov was the first to derive an expression, which led to the spectrum model

$$\Phi_n(\kappa) = 0.033 C_n^2 \kappa^{-11/3}, \tag{6}$$

where κ is the scalar spatial frequency (in rad m^{-1}).

Although Eq. (6) is only valid over th inertial subrange, $1/L_0 \ll \kappa \ll 1/l_0$, it is often assumed that the outer scale is infinite and the inner scale is negligibly small in order to make use of it for all spatial frequencies. However, in practice, making this assumption can lead to untrustworthy results when using the Kolmogorov spectrum for spatial frequencies out of the actual inertial range.

To overcome the singularities appearing in Eq. (6) other spectrum models have been proposed. Tatarskii suggested to include the inner scale effects with a Gaussian function, defining a new power spectral density for refractive-index fluctuations in the form

$$\Phi_n(\kappa) = 0.033 C_n^2 \kappa^{-11/3} \exp\left(-\frac{\kappa^2}{\kappa_m^2}\right), \quad \kappa \gg 1/L_0; \quad \kappa_m = 5.92/l_0. \tag{7}$$

The Tatarskii spectrum still presents a mathematical singularity at $\kappa = 0$ in the limiting case $L_0 \to \infty$. A further improvement of the Tatarskii and Kolmogorov spectrum, valid for all spatial frequencies, called the von Kármán spectrum is given by the expression

$$\Phi_n(\kappa) = 0.033 C_n^2 \frac{\exp(-\kappa^2/\kappa_m^2)}{(\kappa^2 + \kappa_0^2)^{11/3}}, \quad 0 \le \kappa < \infty; \quad \kappa_m = 5.92/l_0, \tag{8}$$

where $\kappa_0 = 2\pi/L_0$.

It should be noted that both Eq. (7) and Eq. (8) reduce to the Kolmogorov power spectrum, when evaluated in the inertial range $\kappa_0 \ll \kappa \ll \kappa_m$.

The spatial power spectral density of refractive-index fluctuations, as being derived from a locally homogeneous random field, is described by its structure function defined by

$$D_n(\vec{r}) = 8\pi \int_0^\infty \kappa^2 \Phi_n(\vec{\kappa}) \left(1 - \frac{\sin \vec{\kappa} \cdot \vec{r}}{\vec{\kappa} \cdot \vec{r}}\right) d\vec{\kappa}, \tag{9}$$

where $\Phi_n(\vec{\kappa})$ is the power spectrum model of interest.

3. Propagation theory

An optical wave propagating through the atmosphere will be altered by refractive-index inhomogeneities that form turbulent eddies of different sizes, where energy is injected in the macroscale L_0 and transfered through ever smaller eddies and finally is dissipated at the microscale l_0. This energy transfer causes unstable air masses, with temperature gradients, giving rise to local changes in the atmospheric refractive-index and thus inducing perturbations as the optical wave propagates.

The random variations on the amplitude and phase of the traveling wave can be addressed theoretical, by solving the wave equation for the electric field and its respective statistical moments. For a propagating electromagnetic wave the electric field is derived from the *stochastic Helmholtz equation*

$$\nabla^2 \vec{E} + k^2 n^2(\vec{r})\vec{E} = 0, \tag{10}$$

where $k = 2\pi/\lambda$ is the wavenumber, $\vec{r}$ is a point in the space and $n(\vec{r})$ is given by Eq. (1).

Traditionally, the actual equation to be solved is the *scalar* stochastic Helmholtz equation

$$\nabla^2 U + k^2 n^2(\vec{r})U = 0, \tag{11}$$

which corresponds to one of three components of the electric field.

To solve Eq. (11) the Born and Rytov approximations have traditionally been used. Additionally, several assumptions are made, namely, backscattering and depolarization effects are neglected, the refractive-index is assumed uncorrelated in the direction of propagation, and the paraxial approximation can be used.

3.1 Born approximation

In the Born approximation the solution of Eq. (11) is assumed to be a sum of terms of the form

$$U(\vec{r}) = U_0(\vec{r}) + U_1(\vec{r}) + U_2(\vec{r}) + \cdots, \tag{12}$$

where $U_0(\vec{r})$ represents the unperturbed field–i.e. an optical wave traveling through free-space. While $U_1(\vec{r})$ and $U_2(\vec{r})$ denote first-order, second-order, and so on, perturbations caused by inhomogeneities due to the random term $n_1(\vec{r})$ in Eq. (1).

Next, by using the fact that in Eq. (1) $n_o \cong 1$ and $|n_1(\vec{r})| \ll 1$, Eq. (11) reduces to

$$\nabla^2 U(\vec{r}) + k^2 \left[1 + 2n_1(\vec{r})\right] U(\vec{r}) = 0, \tag{13}$$

Finally, substituting Eq. (12) into Eq. (13) yields to (Andrews & Philips, 2005)

$$\nabla^2 U_0 + k^2 U_0 = 0, \tag{14}$$

$$\nabla^2 U_1 + k^2 U_1 = -2k^2 n_1(\vec{r})U_0(\vec{r}), \tag{15}$$

$$\nabla^2 U_2 + k^2 U_2 = -2k^2 n_1(\vec{r})U_1(\vec{r}), \tag{16}$$

and so on for higher order perturbations terms.

Solving Eq. (14) gives the unperturbed propagated optical field, whereas solving Eq. (15) and Eq. (16) give the two lower-order perturbed fields. Next, a brief explanation on how to solve this system of equations is given below.

3.1.1 Unperturbed field

Let us refer to Fig. 1 and consider a Gaussian beam wave propagating in the z direction, where the input plane $z = 0$ of the system is located at the output of the laser, and the output plane is located at the receiver lens position. The initial field can be described by (Ishimaru, 1969)

$$U_0(r,0) = A \exp\left(-\frac{r^2}{W_0^2}\right) \exp\left(-i\frac{kr^2}{2F_0}\right), \tag{17}$$

where r is the distance from the beam center, and W_0 and F_0 are the beam radius and phase front radius at the transmitter plane, respectively.[1]

---

[1] The notation used in this section is taken from Andrews & Philips (2005). Special care have to be taken with this notation, where W_0 is specifically referring to the beam radius at the output of the light source, and it should not be confused with the actual beam waist of a Gaussian beam W_B.

Furthermore, the Gaussian beam can be characterized by the input parameters

$$\Theta_0 = 1 - \frac{z}{F_0}, \tag{18}$$

$$\Lambda_0 = \frac{2z}{kW_0^2}, \tag{19}$$

and by the output parameter in the receiver plane at $z = L$

$$\Theta = 1 + \frac{L}{F} = \frac{\Theta_0}{\Theta_0^2 + \Lambda_0^2}, \tag{20}$$

$$\Lambda = \frac{2L}{kW^2} = \frac{\Lambda_0}{\Theta_0^2 + \Lambda_0^2}, \tag{21}$$

where W and F are the beam radius and phase front radius at the receiver plane, respectively.

The set of parameters defining a Gaussian beam presented above correspond to the notation used in Andrews & Philips (2005). Nevertheless, other ways of characterizing a Gaussian beam can be utilized, such that used in Ricklin et al. (2006).

The solution of Eq. (11) for Gaussian beam wave propagating a distance z in free-space is given by

$$U_0(r,z) = \frac{1}{\sqrt{\Theta_0^2 + \Lambda_0^2}} \exp\left(-\frac{r^2}{W^2}\right) \exp\left[i\left(kz - \varphi - \frac{kr^2}{2F}\right)\right], \tag{22}$$

where Θ_0 and Λ_0 are non-dimensional parameters defined above, and φ, W, and F are the longitudinal phase shift, beam radius, and radius of curvature after propagating a distance z. These quantities are defined by

$$\varphi = \tan^{-1}\frac{\Lambda_0}{\Theta_0}, \tag{23}$$

$$W = W_0\sqrt{\Theta_0^2 + \Lambda_0^2}, \tag{24}$$

$$F = \frac{kW_0^2}{2}\left[\frac{\Lambda_0(\Theta_0 + \Lambda_0)}{\Theta_0(1 - \Theta_0) - \Lambda_0}\right]. \tag{25}$$

3.1.2 Perturbations terms

For an optical wave propagating a distance L in the z direction, the first-order perturbation term of the output field is given by

$$U_1(\vec{r}) = 2k^2 \iiint_V G(\vec{r},\vec{s})n_1(\vec{s})U_0(\vec{s})d\vec{s}, \tag{26}$$

where $U_0(\vec{s})$ and $G(\vec{r},\vec{s})$ are the unperturbed field (see Eq. (22)) and the free-space *Green's function* (Yura et al., 1983), respectively. Moreover, by applying the paraxial approximation the first Born approximation reduces to

$$U_1(\vec{r},L) = \frac{k^2}{2\pi}\int_0^L dz \iint_\infty^\infty d^2s \exp\left[ik(L-z) + \frac{ik|\vec{s}-\vec{r}|^2}{2(L-z)}\right] U_0(\vec{s},z)\frac{n_1(\vec{s},z)}{L-z}. \tag{27}$$

When solving for higher-order perturbation terms in the Born approximation, the following recurrent formula can be used

$$U_m(\vec{r}, L) = \frac{k^2}{2\pi} \int_0^L dz \iint_\infty^\infty d^2 s \exp\left[ik(L-z) + \frac{ik|\vec{s}-\vec{r}|^2}{2(L-z)}\right] U_{m-1}(\vec{s}, z) \frac{n_1(\vec{s}, z)}{L-z}, \quad (28)$$

where m indicates the order of the perturbation term to be calculated.

3.2 Rytov approximation

The Rytov approximation assume a solution for Eq. (11) formed by the unperturbed field $U_0(\vec{r})$ modified by complex phase perturbation terms, expressed as

$$U(\vec{r}) = U_0(\vec{r}) \exp\left[\psi_1(\vec{r}) + \psi_2(\vec{r}) + \cdots\right], \quad (29)$$

where $\psi_1(\vec{r})$ and $\psi_2(\vec{r})$ are first- and second-order phase perturbations terms, respectively. These perturbations are defined by (Yura et al., 1983)

$$\begin{aligned} \psi_1(\vec{r}) &= \Phi_1(\vec{r}), \\ \psi_2(\vec{r}) &= \Phi_2(\vec{r}) - \tfrac{1}{2}\Phi_1^2(\vec{r}), \end{aligned} \quad (30)$$

where the new function $\Phi_m(\vec{r})$ appearing in the system of equations in Eq. (30) are directly related with the Born perturbation terms in the form

$$\Phi_m(\vec{r}) = \frac{U_m(\vec{r})}{U_0(\vec{r})} \quad (31)$$

Historically, the Born approximation was first introduced but its results were limited to conditions of extremely weak scintillation. Afterwards, the second-order Rytov approximation won more acceptance thanks to the good agreement with scintillation data in the weak fluctuation regime.

3.3 Statistical moments

The first relevant statistical moment for a traveling optical field is the second-order moment, also known as the *mutual coherence function* (MCF), which is defined as the ensemble average over two points of the field, taken in a plane perpendicular to the propagation direction at a distance L from the source, as follows

$$\Gamma_2(\vec{r_1}, \vec{r_2}, L) = \langle U(\vec{r_1}, L)U^*(\vec{r_2}, L)\rangle, \quad (32)$$

where $U(\vec{r})$ is the Rytov approximation solution for Eq. (11), and the brackets $\langle \cdot \rangle$ denote an ensemble average. Vectors $\vec{r_1}$ and $\vec{r_2}$ are transversal vectors without z component, which is chosen as the propagation direction.

Solving Eq. (32)

$$\Gamma_2(\vec{r_1}, \vec{r_2}, L) = \Gamma_2^0(\vec{r_1}, \vec{r_2}, L) \exp\left[\sigma_r^2(\vec{r_1}, L) + \sigma_r^2(\vec{r_2}, L) - T\right] \exp\left[\frac{-1}{2}\Delta(\vec{r}, \vec{r}, L)\right], \quad (33)$$

where

$$\Gamma_2^0(\vec{r}_1, \vec{r}_2, L) = U_0(\vec{r}_1, L) U_0^*(\vec{r}_2, L) \langle \exp\left[\psi(\vec{r}_1, L) + \psi^*(\vec{r}_2, L)\right]\rangle, \tag{34}$$

$$\sigma_r^2(r, L) = 2\pi^2 k^2 L \int_0^1 \int_0^\infty \kappa \Phi_n(\kappa) \exp\left(-\frac{\Lambda L \kappa^2 \xi^2}{k}\right) [I_0(2\Lambda r \xi \kappa) - 1]\, d\kappa d\xi, \tag{35}$$

$$\sigma_r^2(0, L) = 2\pi^2 k^2 L \int_0^1 \int_0^\infty \kappa \Phi_n(\kappa) \exp\left(-\frac{\Lambda L \kappa^2 \xi^2}{k}\right) \left\{1 - \cos\left[\frac{L\kappa^2}{k}\xi(1 - \overline{\Theta}\xi)\right]\right\} d\kappa d\xi, \tag{36}$$

and $I_0(\cdot)$ is the modified Bessel function of zero order, T is a term denoting the fluctuations of on-axis mean irradiance at the receiver plane caused by atmospheric turbulence (see Andrews & Philips, 2005, Chap. 6.3), and the most-right exponential of Eq. (35) is the complex degree of coherence (DOC).

From the MCF and the DOC physical effects on the optical traveling wave can be derived, namely, the mean irradiance, turbulence-induced beam spreading, angle-of-arrival fluctuations and beam wander.

Actually, the most noticeable effect caused by atmospheric turbulence is the optical scintillation, and it is quantified by the *scintillation index* (SI)

$$\sigma_I^2(\vec{r}, L) = \frac{\langle I^2(\vec{r}, L)\rangle}{\langle I(\vec{r}, L)\rangle^2} - 1, \tag{37}$$

where $I(\vec{r}, L)$ denotes the irradiance of the optical field in the receiver plane.

The mathematical derivation of the SI relies upon the fourth statistical moment of the optical field $U(\vec{r})$, given by

$$\Gamma_4(\vec{r}_1, \vec{r}_2, \vec{r}_3, \vec{r}_4, L) = \langle U(\vec{r}_1, L) U^*(\vec{r}_2, L) U(\vec{r}_3, L) U^*(\vec{r}_4, L)\rangle. \tag{38}$$

By setting $\vec{r}_1 = \vec{r}_2 = \vec{r}_3 = \vec{r}_4 = \vec{r}$ and evaluating Eq. (32) and Eq. (38) for the same point, yields

$$\langle I^2(\vec{r}, L)\rangle = \Gamma_4(\vec{r}, \vec{r}, \vec{r}, \vec{r}, L), \tag{39}$$

$$\langle I(\vec{r}, L)\rangle = \Gamma_2(\vec{r}, \vec{r}, L), \tag{40}$$

thus, obtaining a theoretical expression for the scintillation index.

A fundamental parameter in the study of optical wave propagation through random media is the Rytov variance σ_R^2, which is in fact the scintillation index for a plane wave in the weak turbulence regime. The Rytov variance can be derived from Eq. (36), and by setting $\Lambda = 0$ and $\Theta = 1$ in the limiting case of a plane wave, yields to

$$\sigma_R^2 = 1.23 C_n^2 k^{7/6} L^{11/6}. \tag{41}$$

A more detailed explanation on the derivation of the solution of Eq. (10), and the statistical moments of the optical field can be found in (Andrews & Philips, 2005).

3.4 Extended Rytov theory

The Rytov approximation is valid only in weak irradiance fluctuations regime, and an extension of the theory is needed to address stronger turbulence effects on optical traveling waves. As a wave propagates through the turbulent atmosphere its degree of transverse spatial coherence decreases, this coherence lost is quantified by the *spatial coherence radius*

$$\rho_0 = \begin{cases} \left(\frac{3}{1+\Theta+\Theta^2+\Lambda^2}\right)^{1/2}\left(1.87C_n^2k^2Ll_0^{-1/3}\right)^{-1/2}, & \rho_0 \ll l_0 \\ \left(\frac{8}{3(a+0.62\Lambda^{11/6})}\right)^{3/5}\left(1.46C_n^2k^2L\right)^{-3/5}, & l_0 \ll \rho_0 \ll L_0, \end{cases} \qquad (42)$$

where a is a constant (see Andrews & Philips, 2005, p. 192). It should be noted that Θ and Λ are dimentionless parameters associated with the Gaussian beam. The expression for ρ_0 in the limiting cases of a plane wave ($\Lambda = 0$, $\Theta = 1$), and a spherical wave ($\Lambda = 0$, $\Theta = 0$) can be deduced form Eq. (42).

Another parameter to measure the spatial coherence is the *atmospheric coherence width* $r_0 = 2.1\rho_0$, widely known as the Fried parameter. For the limiting case of a plane wave the Fried parameter is given by

$$r_0 = \left(0.42C_n^2k^2L\right)^{-3/5}. \qquad (43)$$

Under the extended Rytov theory the refractive-index $n_1(\vec{r})$ in Eq. (1) can be seen as the result of the influence of two terms, i.e., the large-scale inhomogeneities $n_X(\vec{r})$ and the small-scale inhomogeneities $n_Y(\vec{r})$. Thus, as the refractive-index directly influences the turbulence power spectrum, an *effective* power spectral density for refractive-index fluctuations can be expressed by

$$\Phi_{ne}(\kappa) = \Phi_n(\kappa)G(\kappa, l_0, L_0) = \Phi_n(\kappa)\left[G_X(\kappa, l_0, L_0) + G_Y(\kappa, l_0)\right], \qquad (44)$$

where G_X and G_Y are amplitude spatial filters modeling the large-scale and small-scale perturbations, respectively.

The effective atmospheric spectrum can be used instead of the classic spectrum to solve the statistical moments of a traveling optical field, thus, allowing to treat the effects of inner-scale size and outer-scale size of turbulence separately throughout the theory.

3.5 Physical effects

3.5.1 Angle-of-arrival fluctuations

Referring to Fig. 1, the rays (solid arrows) leaving the laser source are deflected as they travel through the turbulent atmosphere, some arriving off-axis instead of what is expected without turbulence, represented with the horizontally straight dashed arrow. As the rays may also be interpreted as the wave vector for the traveling wavefront, the variations in the angle respect the optical axis at the receiver represent the concept of angle-of-arrival fluctuations. The expression for the angle-of-arrival fluctuations, that directly depends on the turbulence strength and the optical path length, is given by

$$\langle \beta_a^2 \rangle = 2.91C_n^2L(2W_G)^{-1/3}, \qquad (45)$$

where W_G is soft aperture radius, and it is related to the receiving aperture D by $D^2 = 8W_G^2$.

The main technique to counterbalance the degrading effects of receiving the optical wave off-axis, is by the combination of *fast steering mirrors* and adaptive optics algorithms (Levine et al., 1998; Tyson, 2002; Weyrauch & Vorontsov, 2004).

3.5.2 Beam wander

The beam wander effect is related with the displacement of the instantaneous center of the beam—defined as the point of maximum irradiance—of a traveling wave over the receiver plane. It is well known that this phenomenon is caused by the large-scale inhomogeneities due to their refractive effects. A Gaussian beam wave after propagating through the turbulent atmosphere is corrupted in such a way that the instantaneous field, at the receiver plane, greatly differs of what if expected for a Gaussian beam, with the added characteristic that the beam center can exhibit great deviation from the optical axis of the optical link.

Instead, the short-term and long-term fields have a field shape that resembles that of a Gaussian beam. Nevertheless, the optical field in the short-term exposure is skewed from a Gaussian beam profile, while the long-term profile describes a more accurately Gaussian profile and the deviation of the beam center from the optical axis is relatively small, and can be neglected. A computer simulation of a Gaussian beam, shown in Fig. 4, was conducted following the method described in Dios et al. (2008), where the field profile for different exposure times is presented. For this simulation it was assumed a propagation distance of $L = 2000$m, $C_n^2 = 0.6 \cdot 10^{-14}m^{-2/3}$, $\lambda = 1064$nm, $W_0 = 2$cm, and the exposition time of the long-term profile in Fig. 4(c) is 34 times of that used for the short-term profile in Fig. 4(b).

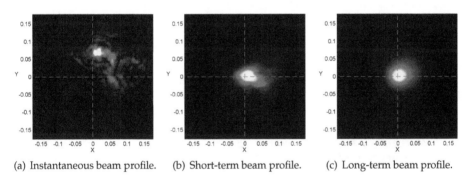

(a) Instantaneous beam profile.   (b) Short-term beam profile.   (c) Long-term beam profile.

Fig. 4. Profile of a Gaussian beam with different exposure times, after propagating .

Fante (1980) in his work relate the random displacements of the incoming wavefront center or "hot spot" with the long-term W_{LT} and short-term W_{ST} spot sizes, assuming that the "hot spot" coincides with beam centroid, by the expression

$$\langle r_c^2 \rangle = W_{LT}^2 - W_{ST}^2, \tag{46}$$

where W_{LT} is the long-term beam radius, and following a conventional Rytov method, its form is (Recolons et al., 2007)

$$W_{LT}^2 = W^2 \left[1 + 1.33\sigma_R^2 \Lambda^{5/6} \right], \quad \Lambda = \frac{2L}{kW^2}, \tag{47}$$

where W is the pure diffraction beam radius at the receiver plane. Furthermore, the short-term beam radius is given by

$$W_{ST}^2 = W^2 \left\{ 1 + 1.33\sigma_R^2 \Lambda^{5/6} \left[1 - 0.66 \left(\frac{\Lambda_0^2}{1+\Lambda_0^2} \right)^{1/6} \right] \right\}. \tag{48}$$

It is clear how in Eq. (47) and Eq. (48) the extra spreading effect due to the atmospheric turbulence is included through the Rytov variance σ_R^2.

For the sake of simplicity, the *geometrical optics approximation* used by Churnside & Lataitis (1990) yields to a closed form expression for the beam wander

$$\langle r_c^2 \rangle = 0.97 C_n^2 L^3 W_0^{-1/3}, \tag{49}$$

while taking into account that this expression is valid for an infinite outer scale and a collimated beam, as is mostly assumed.

3.5.3 Scintillation

A laser beam propagating through the atmosphere will be altered by refractive-index inhomogeneities. At the receiver plane, a random pattern is produced both in time and space (Churnside, 1991). The irradiance fluctuations over the receiver plane resemble the speckle phenomenon observed when a laser beam impinges over a rugged surface. The parameter that express these irradiance fluctuations is the scintillation index

$$\sigma_I^2 = \frac{\langle I^2 \rangle - \langle I \rangle^2}{\langle I \rangle^2} = \frac{\langle I^2 \rangle}{\langle I \rangle^2} - 1, \tag{50}$$

where $I \equiv I(0,L)$ denotes irradiance of the optical wave observed by a point detector after propagating a distance L.

Classical studies on optical wave propagation have been classified in two major categories, either the weak or strong fluctuations theory. It is customary to discriminate both cases for a given propagation problem by determining the value of the Rytov variance. The weak fluctuations regime occurs when $\sigma_R^2 < 1$, the strong fluctuations regime is associates with $\sigma_R^2 > 1$, while there is the saturation regime when $\sigma_R^2 \to \infty$. Different expression are derived for the SI depending on whether the calculation have to be done in the weak or the strong fluctuations regime, although, when $\sigma_R^2 \sim 1$ both expression will give similar results. Andrews et al. (2001) have developed a set of expressions for the SI of Gaussian-beam waves and claimed to be valid in the weak-to-strong fluctuation regime. This work is based on the extended Rytov theory in combination with the solution of the Helmholtz equation given by Eq. (29). The idea behind this approach is to separate the influence of the turbulence in two parts, namely, that caused by the small-scale eddies—that are assumed to be diffractive inhomogeneities—on one hand, and, on the other hand, the effects caused by the large-scale eddies—regarded as refractive inhomogeneities. Mathematically the irradiance is then written as

$$\hat{I} = \frac{I}{\langle I \rangle} = XY, \tag{51}$$

where X and Y are unit mean statistically independent processes arising from the large-scale and small-scale size of turbulence, respectively. Alternatively, the irradiance can be written as $I = A \exp(2\chi)$, where χ is the log-amplitude of the optical wave. Moreover, when χ is normally distributed, the variance of the log-amplitude is related to scintillation index according to

$$\sigma_I^2 = \exp\left(4\sigma_\chi^2\right) - 1 = \exp\left(\sigma_{\ln I}^2\right) - 1, \tag{52}$$

where $\sigma_{\ln I}^2$ is the variance of the log-irradiance, that in turn depends on the large-scale $\sigma_{\ln X}^2$ and small-scale $\sigma_{\ln Y}^2$ variances, as follows

$$\sigma_{\ln I}^2 = 4\sigma_\chi^2 = \sigma_{\ln X}^2 + \sigma_{\ln Y}^2. \tag{53}$$

On the other hand, the on-axis scintillation index for a point receiver takes the integral form

$$\sigma_I^2 = 8\pi^2 k^2 L \int_0^1 \int_0^\infty \kappa \Phi_n(\kappa) \exp\left(-\frac{\Lambda L \kappa^2 \xi^2}{k}\right) \left\{ 1 - \cos\left[\frac{L\kappa^2}{k}\xi(1 - \overline{\Theta}\xi)\right] \right\} d\kappa d\xi \tag{54}$$

where $\overline{\Theta} = 1 - \Theta$, and $\Phi_n(\kappa)$ can be replaced by the effective spectrum in Eq. (44) in order to account for the effects produced by the large-scale G_X and small-scale perturbations G_Y, defined by

$$G_X(\kappa) = \exp\left(\frac{\eta}{\eta_X}\right), \quad \eta = \frac{L\kappa^2}{k},$$

$$G_Y(\kappa) = \frac{\kappa^{11/3}}{(\kappa^2 + \kappa_Y^2)^{11/6}} \exp\left[\frac{\Lambda L \kappa^2 (1 - z/L)^2}{k}\right], \quad \kappa_Y \gg 1, \tag{55}$$

where η_X and κ_Y are variables including the refractive-index structure parameter and some Gaussian beam parameters (see Andrews & Philips, 2005, Sec. 9.6.2). It should be noted that Eq. (55) does not account for inner-scale and outer-scale of turbulence effects.

It is noteworthy that under weak turbulence regime Eq. (52) and Eq. (53) yields to $\sigma_I^2 \cong \sigma_{\ln I}^2 = 4\sigma_\chi^2 = \sigma_{\ln X}^2 + \sigma_{\ln Y}^2$, and following the strategy described above, combining Eq. (44) and Eq. (55) with Eq. (54), the optical scintillation index on-axis of a Gaussian beam for a point receiver is (Andrews et al., 2001)

$$\sigma_I^2 = \exp\left(\sigma_{\ln X}^2 + \sigma_{\ln Y}^2\right) - 1 = \exp\left\{ \frac{0.49\sigma_B^2}{\left[1 + 0.56(1 + \Theta)\sigma_B^{12/5}\right]^{7/6}} + \frac{0.51\sigma_B^2}{\left(1 + 0.69\sigma_B^{12/5}\right)^{5/6}} \right\} - 1. \tag{56}$$

where σ_B^2 is the Rytov variance for a Gaussian beam wave, and it is given by

$$\sigma_B^2 = 3.86\sigma_R^2 \left\{ 0.40[(1 + 2\Theta)^2 + 4\Lambda^2]^{5/12} \cos\left[\frac{5}{6}\tan^{-1}\left(\frac{1 + 2\Theta}{2\Lambda}\right)\right] - \frac{11}{16}\Lambda^{5/6} \right\}, \tag{57}$$

where Θ and Λ are defined by Eq. (20) and Eq. (21), respectively.

On the other hand, the expression of the scintillation index for a receiver over a finite aperture D is given by

$$\sigma_I^2(D) = 8\pi^2 k^2 L \int_0^1 \int_0^\infty \kappa \Phi_n(\kappa) \exp\left(-\frac{L\kappa^2}{k(\Lambda+\Omega_G)}\left[(1-\overline{\Theta}\xi)^2 + \Lambda\Omega_G\xi^2\right]\right)$$

$$\times \left(1 - \cos\left[\frac{L\kappa^2}{k}\left(\frac{\Omega_G-\Lambda}{\Omega_G+\Lambda}\right)\xi(1-\overline{\Theta}\xi)\right]\right) d\kappa d\xi, \quad \Omega_G \geq \Lambda, \tag{58}$$

where $\Omega_G = 2L/kW_G$ is a non-dimensional parameter defining the beam radius at the colleting aperture element. A tractable expression for Eq. (58) have been derived, based on the large-scale and small-scale variances (Andrews & Philips, 2005). Following Eq. (53)

$$\sigma_I^2(D) = \exp\left(\sigma_{\ln X}^2(D) + \sigma_{\ln Y}^2(D)\right) - 1. \tag{59}$$

where large-scale and small-scale log-irradiance variances are given by

$$\sigma_{\ln X}^2(D) = \frac{0.49\left(\frac{\Omega_G-\Lambda}{\Omega_G+\Lambda}\right)^2 \sigma_B^2}{\left[1 + \frac{0.4(2+\Theta)(\sigma_B/\sigma_R)^{12/7}}{(\Omega_G+\Lambda)\left(\frac{1}{3}-\frac{1}{2}\overline{\Theta}+\frac{1}{5}\overline{\Theta}^2\right)^{6/7}} + 0.56(1+\Theta)\sigma_B^{12/5}\right]^{7/6}}, \tag{60}$$

$$\sigma_{\ln Y}^2(D) = \frac{0.51\sigma_B^2(\Omega_G+\Lambda)\left(1+0.69\sigma_B^{12/5}\right)^{-5/6}}{\Omega_G+\Lambda+1.20(\sigma_R/\sigma_B)^{12/5}+0.83\sigma_R^{12/5}}. \tag{61}$$

It should be noticed that these expressions do not account for the inner and outer scale of turbulence effects. In order to include them, additional considerations must be made when developing Eq. (53) as shown in Andrews et al. (2001).

In Fig. 5 a plot of the scintillation index is shown, where different receiving aperture diameters have been used to calculate Eq. (59), and a collimated Gaussian beam have been assumed with wavelength $\lambda = 780$nm and beam size at the transmitter $W_0 = 1.13$cm. Additionally, the effects of the inner scale and the outer scale of turbulence were neglected, and the transmitter aperture diameter was set to 3cm.

From the analysis of the Fig. 5 it can be conclude that a wireless optical communication link can be classified in one of three well differentiated zones. In the first one, regarded as the weak turbulence regime, the scintillation index increases monotonically as either the optical turbulence, denoted by the refractive-index structure parameter, or the link distance increases. Next, a peak in the scintillation index appear representing the point of maximum atmospheric turbulence. This zone is known as the strong turbulence regime. Finally, a third zone called the saturation regime settles the value of the scintillation index to a plateau. The physical reason for the constant level of the SI, irrespective of the increase of the C_n^2 value or the link range, is because after a certain point the atmospheric turbulence completely breaks the spatial coherence of the traveling wavefront and the arriving optical power behaves as a diffuse source. It becomes evident that the localization of the three turbulence regimes explained before is affected by the size of the receiving aperture size.

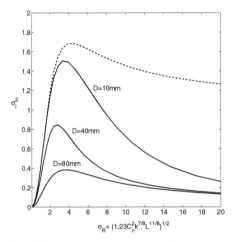

Fig. 5. Scintillation index for different receiver aperture diameters. As reference the SI for a point receiver is shown (dashed line).

3.5.4 Aperture averaging

To counterbalance the scintillation effects on the optical link performance, it is desirable to have a large area at the detection plane in order to integrate as much light as possible. The received wavefront can be regarded as a self-interference pattern, produced by atmospheric inhomogeneities of different spatial scale sizes, that is averaged over the entire receiving area, thus, the intensity fluctuations are mitigated. From the ray optics point of view more rays, which all travel through distinct optical paths, can be collected by means of a lens to be integrated on the photodetector and the measured scintillation index will be lesser compared to that of a point receiver. This phenomenon, called aperture averaging, have been extensively addressed (Andrews, 1992; Churnside, 1991; Fried, 1967). Churnside (1991) developed simple closed expressions to easily evaluate aperture averaging under weak fluctuation regime, that were later corrected by Andrews (1992). More recently, expression for aperture averaging of Gaussian beams have developed for the moderate-to-strong turbulences regime (Andrews et al., 2001; Andrews & Philips, 2005).

The mathematical expression for the aperture averaging factor A is defined by

$$A(D) = \frac{\sigma_I^2(D)}{\sigma_I^2(0)}, \tag{62}$$

where $\sigma_I^2(D)$ is the SI of a receiving aperture with diameter D, and $\sigma_I^2(0)$ is the SI for a point receiver defined in Eq. (56). As far as experimental data is concerned an appropriate way to evaluate the aperture averaging factor is by using an effective point receiver, defined as an aperture much smaller than $\sqrt{\lambda L}$ and the inner scale size l_0 (Ochs & Hill, 1985). It is noteworthy that the lowest possible value of A is desirable, in order to overcome signal fluctuations due to atmospheric turbulence.

In Fig. 6 the aperture averaging factor is shown for Gaussian beam with the same characteristics of that used to plot Fig. 5. Additionally, a link distance of 2km was set. It is clear that for higher turbulence strength the aperture averaging effect becomes less noticeable.

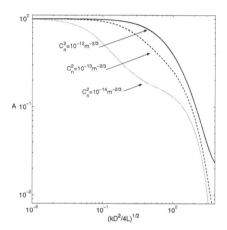

Fig. 6. Aperture averaging factor for different atmospheric turbulence conditions as a function of the receiving aperture radius $D/2$ normalized to the Fresnel zone $\sqrt{L/k}$.

Moreover, it is evident that the averaging capability of the receiver system increases as the the receiving aperture diameter increases.

4. Propagation simulation

Since its introduction by Fleck et al. (1976) the beam split-step method has been widely used to simulate the propagation of electromagnetic waves, where the effects produced by the turbulent atmosphere are simulated by a series of linearly spaced random phase screens. In Fig. 7 are depicted the main aspects involved in the beam split-step method, also known as the *beam propagation method* (BPM). First, an initial traveling optical field is set and the path length L is split into a series of N steps, thus, dividing the optical path into N different slabs of turbulent atmosphere of width $\Delta z = L/N$. Each of this slabs is represented by a two-dimensional (2D) random phase screen placed in the middle of such slab. Consequently, the first and last propagation step have length $\Delta z/2$ while all other steps are Δz in length. The propagation of the optical field between every step takes place in the transformed domain, where the field is decomposed into a linear combination of plane waves. After each step the optical wavefront is inverse transformed, to the spatial domain, where a random phase screen is then used to simulate the atmospheric effects. This process is repeated until the propagation path length is completed.

At the receiver end the detector is a single pixel in the case of a point receiver. When considering a finite size receiving aperture, the optical power in the two-dimensional grid of the traveling wavefront is integrated over the aperture area.

The most widespread technique used to generate the random phase screens is based on the spectral method, in which phase screens are generated in the spectral domain with the selected turbulence power spectrum (Frehlich, 2000; Martin & Flatté, 1998; Recolons & Dios, 2005). The fractal method is an alternative approach to reproduce the phase screens directly in the spatial domain by successive interpolations from a set of random numbers that obey the desired

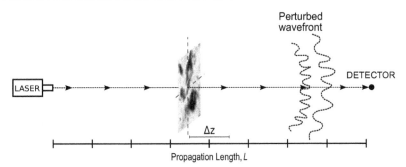

Fig. 7. General scheme of beam propagation method (BPM) applied to the propagation of light through the turbulent atmosphere.

structure function associated with the turbulence power spectrum being used (Lane et al., 1992).

4.1 Spectral method

In this method the screens are numerically generated by the use of fast Fourier-transform and assumed to follow a particular turbulence power spectrum, which most employed is the Kolmogorov spectrum.

The phase screen is generated in the spectral domain by means of filtering Gaussian white noise with the selected turbulence power spectrum, and an inverse transformation yields to the desired random phase screen in the spatial domain, which is given by (Frehlich, 2000)

$$\theta_s(j\Delta x, l\Delta y) = \sum_{n=0}^{N_x} \sum_{m=0}^{N_y} [a(n,m) + ib(n,m)] \times \exp\left[2\pi i (jn/Nx + lm/Ny)\right], \qquad (63)$$

where $i = \sqrt{-1}$; Δx and Δy are the grid spacing; N_x and N_y are the number of points in the respective dimension of the screen; and $a(n,m)$ and $b(n,m)$ are random number following Gaussian white noise statistics with

$$\langle a^2(n,m) \rangle = \langle b^2(n,m) \rangle = \frac{8\pi^3 k^2 \Delta z}{N_x \Delta x N_y \Delta y} \Phi_n(\vec{\kappa}, z), \qquad (64)$$

where k is the wave number, $\Phi_n(\vec{\kappa}, z)$ is the two-dimensional power spectrum for refractive-index fluctuations as a function of the propagation distance z, and $\vec{\kappa}$ is the spatial wave number vector in the plane transversal to propagation direction.

However, a major difficulty with this technique is to reproduce atmospheric large-scale effects owing to the fact that they are related with lowest spatial frequencies of the turbulence spectrum, and, it is precisely around zero where the Kolmogorov spectrum have a singularity. This issue was addressed first by Lane et al. (1992) with the addition of subharmonic components to the random phase screen, as a result of which more resolution in the spatial frequencies around zero is obtained. Later, an improved version of this method was introduced by Recolons & Dios (2005).

4.2 Fractal method

Phase screens generated following a Kolmogorov power spectrum have an important property, namely, that they present a fractal behavior as they look similar regardless of the scale they are viewed. The first to propose the use of fractal interpolation for generating phase screens was Lane et al. (1992), and later an improved version of this method was introduced (Harding et al., 1999).

With this method, first, an exact low-resolution phase screen is generated by evaluating its covariance matrix that is obtained directly from the structure function. Which for a pure Kolmogorov spectrum, as it is normally assumed, is given by

$$D(\vec{r_1}, \vec{r_2}) = \langle[\theta_s(\vec{r_1}) - \theta_s(\vec{r_2})]^2\rangle = 6.88 \left(\frac{|\vec{r_1} - \vec{r_2}|}{r_0}\right)^{5/3}, \tag{65}$$

where $\theta_s(\vec{r_1})$ and $\theta_s(\vec{r_2})$ are the phase evaluated at positions $\vec{r_1}$ and $\vec{r_2}$, respectively, and r_0 is the Fried parameter. The covariance matrix can be obtained from the structure function with the relationship

$$
\begin{aligned}
C_{\theta_s}(\vec{r_1}, \vec{r_2}) = &-\frac{1}{2}D(\vec{r_1}, \vec{r_2}) + \frac{1}{2}\iint D(\vec{r_1}', \vec{r_2})T(\vec{r_1}')dx_1'dy_1' \\
&+\frac{1}{2}\iint D(\vec{r_1}, \vec{r_2}')T(\vec{r_2}')dx_2'dy_2' \\
&-\frac{1}{2}\iiiint D(\vec{r_1}', \vec{r_2}')T(\vec{r_1}')T(\vec{r_2}')dx_1'dy_1'dx_2'dy_2',
\end{aligned}
\tag{66}
$$

where $T(\vec{r})$ is a windowing function that has a constant value inside the dominion of the phase screens and zero value outside. Next, once the covariance matrix is obtained, an square matrix matching the size of the phase screen is generated from a set of Gaussian random numbers with variance given by the eigenvalues of $C_{\theta_s}(\vec{r_1}, \vec{r_2})$ (Harding et al., 1999). Nevertheless, is important to note that if the initial squared phase screen have a size $N \times N$, then the covariance matrix has $N^2 \times N^2$ dimension, making the method applicable to small values of N. Thereby, the use of interpolation techniques rises as mandatory to obtain phase screen with higher resolution. Probably, the most widespread window sizes in the literature are $N = 512$ and $N = 1024$.

When the low-resolution phase screen have been completely generated, successive randomized interpolation steps are executed to produce the desired grid size. This interpolation method achieve a high resolution degree while demanding a relatively small computational effort, although, having the drawback of poorer statistical performance.

5. Wireless optical communication systems

Previous sections have been focused on the explanation of the physical phenomena that affects an optical traveling wave in a free-space optical link, as shown in Fig. 1. From a communication system approach, there are other factors that become critical when evaluating the performance of a wireless optical communications link. A simplified scheme is shown in Fig. 8, where the main factors involved in are presented.

Wireless optical communications rely on a traveling wave generated by a laser source, at certain average power level transmitted P_T. Aside from the effects suffered by the optical

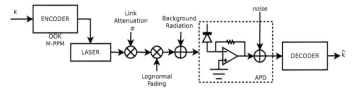

Fig. 8. Block diagram for a wireless optical communication link.

traveling wavefront through the turbulent atmospheric channel, addressed in Section 3, the average optical power at the receiver plane P_R is influenced by various parameter. The expression for the average optical power detected at a distance R in a WOC link, is given by

$$P_R(R) = P_T \frac{D_R^2}{D_T^2 + (R\theta)^2} \exp\left(-\frac{\theta_{mp}^2}{(\theta/2)^2}\right) T_a(R) T_R, \qquad (67)$$

where θ is the laser beam full-angle divergence, $T_a(R)$ is the transmittance of the atmosphere along the optical path, T_R is the transmittance of the receiver optics, θ_{mp} denotes pointing errors between the emitter and receiver, and, D_T and D_R are the transmitting and receiving apertures diameters, respectively. It should be noted that pointing errors not only are due to misalignments in the installation process, but also to vibrations on the transmitter and receiver platforms. For horizontal links the vibration come from transceiver stage oscillations and buildings oscillations caused by wind, while for vertical links—i.e. ground to satellite link— satellite wobbling oscillation are the main source of pointing errors.

5.1 Atmospheric attenuation

A laser beam traveling through the turbulent atmosphere is affected by extinction due to aerosols and molecules suspended in the air. The transmittance of the atmosphere can be expressed by Beer's law as

$$T_a(R) = \frac{P(R)}{P(0)} = e^{-\alpha_a R}, \qquad (68)$$

where $P(0)$ is transmitted laser power at the source, and $P(R)$ is the laser power at a distance R. The total extinction coefficient per unit length α_a comprises four different phenomena, namely, molecular and aerosol scattering , and, molecular and aerosol absorption:

$$\alpha_a = \alpha_{abs}^{mol} + \alpha_{abs}^{aer} + \beta_{sca}^{mol} + \beta_{sca}^{aer}. \qquad (69)$$

The molecular and aerosol behavior for the scattering and absorption process is wavelength dependent, thus, creating atmospheric windows where the transmission of optical wireless signal is more favored. The spectral transmittance of the atmosphere is presented in Fig. 9, for a horizontal path of nearly 2 km at sea level (Hudson, 1969).

Within the atmospheric transmittance windows the molecular and aerosol absorption can be neglected. Molecular scattering is very small in the near-infrared, due to dependence on λ^{-4}, and can also be neglected. Therefore, aerosol scattering becomes the dominating factor reducing the total extinction coefficient to (Kim et al., 1998)

$$\alpha_a = \beta_{sca}^{aer} = \frac{3.91}{V}\left(\frac{\lambda}{550}\right)^{-q}, \qquad (70)$$

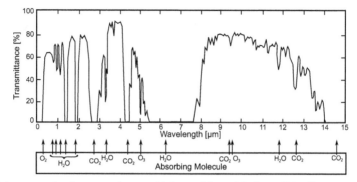

Fig. 9. Earth's atmospheric transmittance [Adapted from Hudson (1969)].

where V is the visibility in kilometers, λ is the wavelength in nanometers, and q is the size distribution of the scattering particles. Typical values for q are given in Table 3.

The attenuation factors that supposed the larger penalties are the atmospheric attenuation and the geometrical spreading losses, both represented in Fig. 10. It becomes evident from the inspection of their respective behaviors, that the atmospheric attenuation imposes larger attenuation factors for poor visibility conditions than the geometrical losses due to the beam divergence of the laser source. Meteorological phenomena as snow and haze are the worst obstacle to set horizontal optical links, and, of course, the clouds in vertical ground-to-satellite links, which imposed the need of privileged locations for deploying optical ground stations.

Visibility	q
$V >50$km	1.6
6km$< V <$50km	1.3
$V <6$km	$0.585V^{1/3}$

Table 3. Value of the size distribution of the scattering particles q, for different visibility conditions.

For the calculations in Fig. 10(a) a light source with wavelength $\lambda = 780$nm was assumed, and for Fig. 10(b) the aperture diameter in transmission and reception was set to 4cm and 15cm, respectively. The negative values of the attenuation in Fig. 10(b) imply that the geometrical spreading, of the transmitted beam, have not yet exceeded in size the receiving aperture.

5.2 Background radiance

In a wireless optical communication link the receiver photodetector is always subject to an impinging optical power, even when no laser pulse have been transmitted. This is because the sun radiation is scattered by the atmosphere, the Earth's surface, buildings, clouds, and water masses, forming a background optical power. The amount of background radiance detected in the receiver depends on the area and the field of view of the collecting telescope, the optical bandwidth of the photodetector, and weather conditions. The most straightforward method to decrease background radiation is by adding an interference filter with the smallest possible optical bandwidth, and the center wavelength matching that of the laser source. Typical values of optical bandwidth these filters are in orders of a few nanometers.

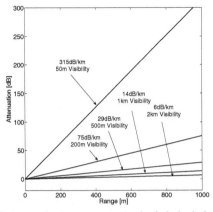 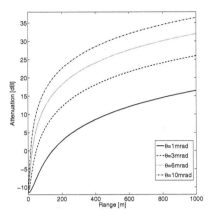

(a) Atmospheric attenuation in decibels for light source with $\lambda = 780$nm and various visibility conditions.

(b) Geometrical spreading loss in decibels for various beam full-angle divergences of light source.

Fig. 10. Attenuation factor dependence on link distance.

The total background radiation can be characterized by the spectral radiance of the sky, which is different for day or night operation. The curves for daytime conditions will be very similar to those of nighttime, with the addition of scattered sun radiation below $3\,\mu$m (Hudson, 1969). The typical behavior of the spectral radiance of the sky is shown in Fig. 11 for daytime condition and an horizontal path at noon.

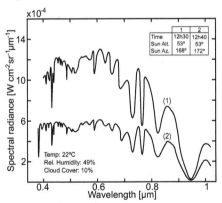

Fig. 11. Spectral radiance of the sky for a clear daytime [Adapted from Knestrick & Curcio (1967)].

Once the spectral radiance of the sky is known the total optical power at the receiver, due to background, can be calculated by

$$P_B = N_B T_R \left(\pi \frac{D}{2} \frac{\text{FOV}}{2} \right)^2 B_{opt}, \tag{71}$$

where N_B is the background spectral radiance, FOV is the field of view of the receiving telescope, and B_{opt} is the optical bandwidth of the interference filter.

Following the method described by Bird & Riordan (1986) an estimation of the diffuse irradiance, considering rural environment, for 830 nm would be between 60 and 100 W m^{-2} μm^{-1}, depending on the elevation angle of the Sun during the day. These values respond to the irradiance received on the ground coming from the sky in all directions without considering the solar crown, and are about the same order of the values presented in Fig. 11. Therefore, special care have to be taken from having direct sun light into the telescope field of view, situation that may produce link outages due to saturation of the photodetector.

5.3 Probability density functions for the received optical power

In any communication system the performance characterization is, traditionally, done by evaluating link parameter such as probability of detection, probability of miss and false alarm; threshold level for a hard-decoder and fade probability, that demands knowledge of the probability density function (PDF) for the received optical power (Wayne et al., 2010). Actually, it is rather a difficult task to determine what is the exact PDF that fits the statistics of the optical power received through an atmospheric path.

Historically, many PDF distributions have be proposed to described the random fading events of the signal-carrying optical beam, leading to power losses and eventually to complete outages. The most widely accepted distributions are the Log-Normal (LN) and the Gamma-Gamma (GG) models, although, many others have been subject of studies, namely, the K, Gamma, exponential, I-K and Lognormal-Rician distributions (Churnside & Frehlich, 1989; Epple, 2010; Vetelino et al., 2007).

In literature, although not always mentioned, the PDF distribution for the received optical power in a wireless link will be greatly influenced whether the receiver have a collecting aperture or it is just the bear photodetector, i.e., a point receiver. Experimental studies support the fact that the LN model is valid in weak turbulence regime for a point receiver and in all regimes of turbulence for aperture averaged data (Perlot & Fritzsche, 2004; Vetelino et al., 2007). On the other hand, the GG model is accepted to be valid in all turbulence regimes for a point receiver, nevertheless, this does not hold when aperture averaging takes place (Al-Habash et al., 2001; Vetelino et al., 2007).

The Log-Normal distribution is given by

$$f_{LN}(I; \mu_{\ln I}, \sigma_{\ln I}^2) = \frac{1}{I \sqrt{2\pi \sigma_{\ln I}^2}} \exp \left\{ -\frac{[\ln(I) - \mu_{\ln I}]^2}{2 \sigma_{\ln I}^2} \right\}, \quad I > 0, \tag{72}$$

where $\mu_{\ln I}$ is the mean and $\sigma_{\ln I}^2$ is the variance of the log-irradiance, and they are related to the scintillation index σ_I^2 by

$$\mu_{\ln I} = \ln\langle I \rangle - \frac{\sigma_{\ln I}^2}{2}, \tag{73}$$

$$\sigma_{\ln I}^2 = \ln \left(\sigma_I^2 + 1 \right). \tag{74}$$

The Gamma-Gamma distribution is used to model the two independent contributions of the small-scale and large-scale of turbulence, assuming each of them is governed by a Gamma process. The GG distribution is given by

$$f_{GG}(I; \alpha, \beta) = \frac{2(\alpha\beta)^{(\alpha+\beta)/2}}{\Gamma(\alpha)\Gamma(\beta)} I^{(\alpha+\beta)/2-1} K_{\alpha-\beta}\left(2\sqrt{\alpha\beta I}\right), \quad I > 0, \tag{75}$$

where $K_n(x)$ is the modified Bessel function of the second kind and order n, and, α and β are parameters directly related to the effects induced by the large-scale and small-scale scattering, respectively (Epple, 2010). The parameters α and β are related to the scintillation index by

$$\sigma_I^2 = \frac{1}{\alpha} + \frac{1}{\beta} + \frac{1}{\alpha\beta}. \tag{76}$$

It is customary to normalize Eq. (72) and Eq. (75) in the sense that $\langle I \rangle = 1$. Under such assumption, the parameters α and β of the GG distribution can be related to the small-scale and large-scale scintillation, introduced in Section 3.5.3, in the form (Andrews & Philips, 2005)

$$\alpha = \frac{1}{\sigma_X^2(D)} = \frac{1}{\exp(\sigma_{\ln X}^2(D)) - 1} \tag{77}$$

$$\beta = \frac{1}{\sigma_Y^2(D)} = \frac{1}{\exp(\sigma_{\ln Y}^2(D)) - 1}. \tag{78}$$

6. Wireless optical communication system performance

6.1 Intensity-modulation direct-detection

One of the fundamental technical decisions for a wireless optical communication systems is the choice of the modulation scheme. Although, many modulation techniques have been proposed, from non-coherent to coherent schemes, there is marked trend to favor the use of intensity-modulation direct-detection (IM/DD) scheme because it hides the high-frequency nature of the optical carrier thanks to its equivalent baseband model (Barry, 1994). Another reason to prefer the IM/DD scheme is the relatively low design complexity of the receiver system, when comparing to coherent systems, and because the photodetector is many times larger than the optical wavelength it exhibits a high degree of immunity to multipath fading (Wong et al., 2000). Among the most widespread intensity-modulation schemes for optical communications, namely on-off keying (OOK) and pulse position modulation (PPM), Chan (1982) noted that the PPM scheme is the most suitable for FSO owing to the fact that it does not rely on a threshold value to apply optimal detection.

The PPM format encodes L bits of information into one symbol, or word, of duration T_w that is divided in $M = 2^L$ slots, by transmitting power in only one out of the M possible slots. Therefore, PPM presents itself as an orthogonal modulation scheme. On the receiver side the maximum-likelihood detection is done by choosing the slot that contains the maximum count of photons—i.e. energy—over a word time, after synchronization have been achieved. The waveform for a set of bits encoded in 8-PPM before and after propagation is shown in Fig. 12.

A communication system based on M-PPM modulation groups the input bits, at the transmitter, in blocks of length $L = \log_2 M$, with bit rate R_b, to transmit them at a symbol rate of $R_w = 1/T_w = R_b/\log_2 M$, where T_w is the word or symbol time. Hence, the bandwidth

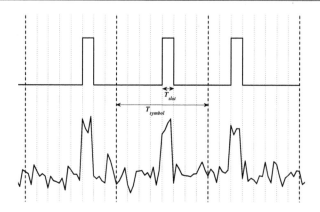

Fig. 12. Binary sequence 101100010 encoded in 8-PPM (top) and same sequence corrupted by noise (bottom).

required for transmitting with M-PPM can be approximated by the bandwidth of a single slot pulse (Barry, 1994)

$$B \approx \frac{1}{T_p} = \frac{1}{T_w/M} = \frac{MR_b}{\log_2 M}, \tag{79}$$

where T_p is the pulse or slot time, and corresponds to a higher requirement in bandwidth compared to that of an OOK modulation scheme, which is $B = R_b$. Table 4 shows the average power and bandwidth requirements for OOK and PPM modulation. The binary phase-shift keying and quadrature phase-shift keying with N subcarriers modulation schemes—BPSK and N-QPSK, respectively—are presented as reference.

Modulation Scheme	Average power	Bandwidth
OOK	$P_{OOK} = \sqrt{N_0 R_b} erf^{-1}(BER)$	R_b
M-PPM	$\dfrac{1}{\sqrt{0.5M \log_2 M}} P_{OOK}$	$\dfrac{MR_b}{\log_2 M}$
BPSK	$\sqrt{2} P_{OOK}$	$2R_b$
N-QPSK	$\sqrt{2N} P_{OOK}$	R_b

Table 4. Average power and bandwidth requirements for different modulation schemes for a given bit-error rate (BER) [Adapted from Barry (1994)].

On the other hand, regarding the average optical power required for achieving a certain bit-error rate (BER), a PPM waveform need $1/\sqrt{0.5M \log_2 M}$ less power than for achieving that same BER with OOK. Consequently, for a given bit rate and BER value PPM modulation demands higher bandwidth and less average optical power, when comparing with OOK. Except for the special case of 2-PPM, where the power requirement is exactly the same as for OOK, while the bandwidth is double. Furthermore, many authors have proposed new modulations derived from PPM, in particular differential PPM (DPPM) (Shiu & Kahn, 1999), overlapping PPM (OPPM) (Patarasen & Georghlades, 1992), improved PPM (IPPM) (Perez-Jimenez et al., 1996) and multipulse PPM (MPPM) (Sigiyama & Nosu, 1989), aiming to overcome the excessive bandwidth requirements of PPM modulation.

Demodulating a PPM signal can be done either by *hard-decoding* or *soft-decoding*. The latter is preferred as it requires, after slot and symbol synchronization have been achieved, to integrate the power on each slot within a frame and then choose the largest. This way the PPM is highly resistance to background noise and the receiver sensibility is increased respect to the hard-decoding approach, where a simple threshold detector is used to decide if a "low" or "high" have been received.

As the information is conveyed in time for PPM modulation format, a critical issue is the synchronization procedure. Timing recovery of the slot and symbol clock become essential to correctly decode the received noisy waveform. Many strategies have been proposed to aid in the synchronization stage (Georghiades & Snyder, 1984; Srinivasan et al., 2005; Sun & Davidson, 1990). Moreover, variants of PPM have been introduced, such as half-pulse PPM (Otte et al., 1998) and digital pulse interval modulation (DPIM) (Okazaki, 1978), in order to simplify the synchronization process.

6.2 Signal-to-noise ratio in an APD photodetector

Several terms of noise have to be taken into account for the evaluation of the signal-to-noise ratio (SNR) in an optical link. Some of them are characteristic of the photo-detector device, as the noise associated to the dark current or the noise coming from the intrinsic gain physical mechanism. Other terms come from the amplifier electronics.

The usual expression for the SNR at the receiver output is

$$\text{SNR} = \frac{i_S^2}{\sigma_S^2 + \sigma_B^2 + \sigma_D^2 + \sigma_A^2 + i_n^2} \tag{80}$$

being i_S^2 the generated photocurrent, σ_S^2 the shot noise associated to the received signal, σ_D^2 the dark current noise, σ_B^2 the noise coming from the background optical power, i_n^2 the thermal noise and σ_A^2 the total equivalent noise input current associated to the amplifier. The photocurrent is calculated as follows

$$i_S = \eta e G \frac{\lambda P_S}{hc} \tag{81}$$

where η is the quantum efficiency, P_S is the received signal optical power, h is the Planck's constant, e is the electron charge, c is the speed of light in vacuum and G is the photo-detector intrinsic gain. The photocurrent can also be expressed by

$$i_S = R_I P_S, \tag{82}$$

where R_I is the current responsivity,

$$R_I = \eta e G \frac{\lambda}{hc} \quad \left[\frac{\text{A}}{\text{W}} \right]. \tag{83}$$

6.2.1 Noise sources

To complete the characterization of the signal-to-noise ratio an analysis of the noise sources in an avalanche photodiode (APD) have to be done. Some noise terms depend directly on the photodetector physical characteristics, while others are generated by the optical power illuminating the surface of the APD.

The process that detects the optical power impinging the detector is described by the occurrence of independent random events, modeled by the Poisson distribution, as an optical wave is ultimately formed by *photons* carrying quantized amounts of energy. This randomness in the detection process of any photon illumination the APD, is what gives rise to the *shot noise* and *background noise*. The shot noise is given by

$$\sigma_S^2 = 2ei_S BGF = 2e(R_I P_S)BGF, \tag{84}$$

and the background noise by

$$\sigma_B^2 = 2ei_B BGF = 2e(R_I P_B)BGF, \tag{85}$$

where B is the electrical bandwidth, P_B is the background optical power, and F the excess noise factor. For a positive-intrinsic-negative (PIN) photodiode this excess noise factor is the unity, as no internal gain exists. For an APD the general expression is

$$F = k_{eff}G + (1 - k_{eff})\left(2 - \frac{1}{G}\right), \tag{86}$$

where k_{eff} is the carrier ionization ratio, for which 0.01-0.1 is the typical range of values.

Every photodetector, whether is a PIN diode or an APD, generates a drift current even when no photons are entering the detector surface. This phenomena is due to the random generation of electron-holes pairs within the depletion area, and charges are attracted by the electric field produced by polarization voltage. The dark current is the result of two current terms, and it is defined by

$$i_D = i_{DS} + i_{DB}G, \tag{87}$$

where i_{DS} is the surface leakage current, and i_{DB} the bulk noise current (for gain unity). The second term is a function of the gain, as it is affected by the avalanche process. Then, the corresponding noise term is

$$\sigma_D^2 = 2e\left(i_{DS} + i_{DB}G^2F\right)B. \tag{88}$$

In an actual system the photodetector is always followed by an amplifier in order to adequate the signal to next stages, in the receiver chain. The amplifier noise is characterized by means of two noisy sources at the input, namely, an equivalent noise voltage source and an equivalent noise current source. The values of these sources appears as two parameters of the amplifier in the datasheet provided by the manufacturer, most of times even with plots showing their behaviour as a function of the modulated signal frequency. In a first approximation the total equivalent noise current of the amplifier can be written as

$$\sigma_A^2 = \left(i_{nA}^2 + \frac{e_{nA}^2}{R^2}\right)B, \tag{89}$$

being R the feedback resistor in the transimpedance amplifier scheme, i_{nA} the amplifier equivalent noise current density, in $A/\sqrt{Hz}$, and e_{nA} the amplifier equivalent noise voltage density, in $V/\sqrt{Hz}$.

Finally, as in any electronic system there will always exist thermal noise, defined by

$$i_{nR}^2 = \frac{4KTB}{R},$$ (90)

where K is the Boltzmann's constant and T is the absolute temperature in Kelvins.

Traditionally the manufacturer gives a characteristic noise figure of the photo-detector known as *noise equivalent power* (NEP). This is defined as the minimum optical signal that could be detected by the device, where *minimum signal* is understood as the signal power level for which the SNR equals unity. For a bared APD the NEP is limited by the dark current, whereas for a complete photodetector module other terms must be included, namely, the electronic and the thermal noise. Moreover, it is a common practice to give that parameter for 1 Hz bandwidth, i.e., normalized with respect to the bandwidth, which may vary from one application to other. The NEP for a complete photodetector (APD plus preamplifier) is given by

$$\text{NEP} = \left.\frac{P_S}{\sqrt{B}}\right|_{\text{SNR}=1} = \frac{1}{R_I}\sqrt{\frac{1}{B}(\sigma_D^2 + \sigma_A^2 + i_n^2)} \quad \left[\frac{W}{\sqrt{Hz}}\right].$$ (91)

By using this merit figure the evaluation of the noise could be abbreviated, as there is no need to calculate all of the terms of noise involved in it. Therefore, the SNR calculation only implies the knowledge of the signal and background power, along with the photodetector NEP, thus

$$\text{SNR} = \frac{i_s^2}{\sigma_s^2 + \sigma_B^2 + R_I^2\text{NEP}^2B}.$$ (92)

6.3 Bit-error rate performance

In digital communication systems, reliability is commonly expressed as the probability of bit error, best known as bit-error rate (BER), measured at the output of the receiver and depends directly on the received signal level and receiver noise level. The smaller the BER, the more reliable the communication system .

In order to obtain an accurate calculation of the BER, it is necessary to know the probability density function of the receiver output signal. In the case of WOC systems the APD is the preferred choice as photodetector. The output of an APD is modeled by the McIntyre-Conradi distribution (Webb et al., 1974), although, the Gaussian approximation is sufficient enough when the bulk current is of the order of nanoampere and the absorbed photons are more than a few hundred within the observation time (Davidson & Sun, 1988; Ricklin et al., 2004).

For the output of a wireless optical communication link, with an APD as photodetector, there are two possibilities, namely, that a pulse is transmitted or not. In the former situation the APD is detecting optical power corresponding to the signal level and the background radiation, while in the latter only background radiation is received. Assuming a Gaussian distribution given by

$$f(x; \mu, \sigma^2) = \frac{1}{\sqrt{2\pi\sigma^2}} \exp\left[-\frac{(x - \mu)^2}{2\sigma^2}\right],$$ (93)

the average current μ_1 and its associated current noise σ_1^2 generated at the APD's output when a pulse have been transmitted is

$$\mu_1 = eG\frac{\eta}{h\nu}(P_S + P_{BG}) + i_{DS} + i_{DB}G, \tag{94}$$

$$\sigma_1^2 = 2B\left(\frac{e^2\eta}{h\nu}FG^2(P_S + P_{BG}) + \frac{2KT}{R_L} + e(i_{DS} + i_{DB}G^2F)\right). \tag{95}$$

On the other hand, when no pulse is transmitted the average current μ_0 and its associated current noise σ_0^2 is

$$\mu_0 = eG\frac{\eta}{h\nu}(\epsilon P_S + P_{BG}) + i_{DS} + i_{DB}G, \tag{96}$$

$$\sigma_0^2 = 2B\left(\frac{e^2\eta}{h\nu}FG^2(\epsilon P_S + P_{BG}) + \frac{2KT}{R_L} + e(i_{DS} + i_{DB}G^2F)\right), \tag{97}$$

where ϵ denotes the laser extinction ratio, generating residual light even when no pulse is being transmitted. All other parameters in Eq. (94) to Eq. (97) were presented in Section 6.2.1.

6.3.1 Probability of error for on-off keying modulation

The simplest signaling format in a digital wireless optical communication system is the on-off keying (OOK), where a binary '1' is represented by a pulse while a binary '0' is represented by the absence of a pulse. The receiver, in this case, is comprised of a threshold detector for deciding which symbol have been received. Assuming that the receiver output noise follows a white Gaussian model, the corresponding PDFs for the cases of a pulse and no pulse being transmitted are shown in Fig. 13, where τ represents the threshold level applied for comparison, and, the mean and variances are defined in Eq. (94) to Eq. (97). Let us consider

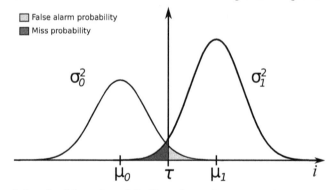

Fig. 13. Received signal p.d.f. under white Gaussian noise assumption.

now the word-error probability (PWE), which is compound of two types of errors. If the received signal level is higher than the set threshold τ when no pulse has been transmitted by the source, a false alarm is generated. On the contrary, if a pulse indeed has been transmitted and the received signal level is lower than τ, then, a miss is produced. Thus, the PWE is given by

$$\text{PWE} = P_{\text{FA}} P_0 + P_{\text{Miss}} P_1, \tag{98}$$

$$P_{\text{FA}} = Q\left(\frac{\tau - \mu_0}{\sigma_0}\right), \tag{99}$$

$$P_{\text{Miss}} = 1 - Q\left(\frac{\tau - \mu_1}{\sigma_1}\right) = Q\left(\frac{\mu_1 - \tau}{\sigma_1}\right), \tag{100}$$

where P_{FA} denotes the probability of false alarm, P_{Miss} is the probability of miss, and $Q(x)$ is the Gaussian Q-function defined as

$$Q(x) = \frac{1}{2\pi} \int_x^\infty e^{-x^2/2} dx. \tag{101}$$

Whenever an equiprobable signaling system is used, the probability of receiving a pulse or not are equal, this is $P_0 = P_1 = \frac{1}{2}$. For OOK modulation, the bit-error probability P_b is the same as the word-error probability—i.e. $P_b = \text{PWE}$.

The problem of defining the optimum threshold level have been addressed before, and the expression for τ in a maximum-likelihood receiver yields to (Ricklin et al., 2004)

$$\left(\frac{\sigma_1^2}{\sigma_0^2} - 1\right) \tau^2 + 2\left(\mu_1 - \frac{\sigma_1^2}{\sigma_0^2}\mu_0\right)\tau - \sigma_1^2 \ln\left(\frac{\sigma_1^2}{\sigma_0^2}\right) + \frac{\sigma_1^2}{\sigma_0^2}\mu_0^2 - \mu_1^2 = 0. \tag{102}$$

Nevertheless, as real-time calculation of the mean and variance of the received signal is rather a complex task, a reasonable approach is to set the threshold level to half of the signal amplitude, which actually approaches the optimum value of τ for high SNR values.

6.3.2 Probability of error for M-ary pulse position modulation

Pulse position modulation is a signaling format well suited for laser applications, requiring low average power and is very resistant to background radiation. In M-ary PPM signaling, L binary source bits are transmitted as a single light pulse in one out of $M = 2^L$ possible time slots, once every T_w seconds.

A maximum-likelihood APD based receiver, with M-PPM modulation, have a word-error probability given by (Dolinar et al., 2006)

$$\text{PWE} = 1 - \int_{-\infty}^{\infty} \sqrt{\frac{\gamma}{\beta + \gamma}} \phi\left(\sqrt{\frac{\gamma}{\beta + \gamma}}(x - \sqrt{\beta})\right) \Phi(x)^{M-1} dx, \tag{103}$$

where $\phi(x)$ is given by Eq. (93) with zero mean and unitary variance, $\Phi(x)$ is the standard Gaussian cumulative distribution function, $\beta = (\mu_1 - \mu_0)^2/\sigma_0^2$ is the symbol signal-to-noise ratio, and, $\gamma = (\mu_1 - \mu_0)^2)/(\sigma_1^2 - \sigma_0^2)$.

In a different approach, a threshold detector can be implemented for demodulating PPM signals. Although, it is not the optimum strategy it can greatly simplify receiver design, as tight synchronization requirements have not to be pursued as for the optimum receiver.

The expression for the word-error probability for a threshold receiver have been derive by Moreira et al. (1996), leading to

$$\text{PWE} = 1 - \left[P_1 + \frac{1}{M}P_2 + \sum_{n=2}^{M} \frac{1}{n}P_{3n} \right], \tag{104}$$

where P_1 is the probability of detecting a pulse in the correct position, P_2 is the probability of that no pulse is detected and P_{3n} is the probability of detecting n pulses. These probabilities are defined by

$$P_1 = (1 - P_{\text{Miss}})(1 - P_{\text{FA}})^{M-1},$$

$$P_2 = P_{\text{Miss}}(1 - P_{\text{Miss}})^{M-1}, \tag{105}$$

$$P_{3n} = \binom{M-1}{n-1}(1 - P_{\text{Miss}})P_{\text{FA}}^{n-1}(1 - P_{\text{FA}})^{M-1},$$

where P_{FA} and P_{Miss} are given by Eq. (99) and Eq. (100), respectively.

Sometimes having the error probability at bit level is desirable. Thus, for a M-ary orthogonal signaling system, the probability of word error can be converted to bit-error probability according to (Proakis, 2001)

$$P_b = \frac{M/2}{M-1}\text{PWE}. \tag{106}$$

6.4 Bit-error rate under turbulent atmosphere

In the presence of optical turbulence, the probability of error is a conditional probability owing to the random nature of the received optical power. Thus, the SNR becomes a random variable and consequently the PWE have to be averaged over all the possible received optical signal levels, according to the proper statistical distribution model of the irradiance I. This yields to

$$\text{PWE}(\sigma_I^2) = \int_0^{\infty} P(I)\text{PWE}(I)dI, \qquad I > 0, \tag{107}$$

where σ_I^2 is the scintillation index defined in Section 3, which depends directly on link's parameters such as C_n^2, link distance, laser divergence, and aperture averaging among others.

As stated in Section 5.3 the Log-Normal model is the most accepted in weak turbulence regimes for point receivers, and in all regimes of turbulence when aperture averaging takes place. Under weak turbulence regime the scintillation index, given by Eq. (52), can be expressed as

$$\sigma_I^2 \cong \sigma_{\ln I}^2, \tag{108}$$

and normalizing the irradiance in the sense that $\langle I \rangle = 1$, Eq. (72) reduces to

$$P(I) = \frac{1}{I\sqrt{2\pi\sigma_I^2}} \exp\left[-\frac{\left(\ln I + \frac{\sigma_I^2}{2}\right)^2}{2\sigma_I^2} \right]. \tag{109}$$

Some examples of the performance of a wireless optical communication system, with

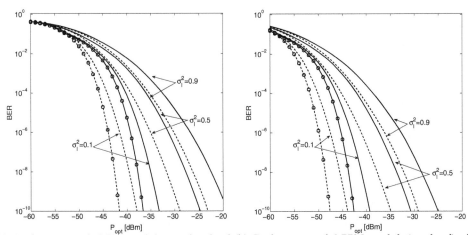

(a) Performance of OOK modulation for fixed threshold (solid line) and optimum threshold (dashed lines) receivers under different turbulence levels. Lines with circles represent free-space (no-turbulence) conditions.

(b) Performance of 8-PPM modulation for fixed threshold (solid line) and optimum threshold (dashed lines) receivers under different turbulence levels. Lines with circles represent free-space (no-turbulence) conditions.

Fig. 14. Performance of OOK and 8-PPM modulation for bit-error rate vs received average optical power in a WOC systems, with IM/DD, for different levels of scintillation index.

intensity-modulation and direct-detection, for OOK and PPM modulation with fixed and optimum threshold are shown in Fig. 14(a) and Fig. 14(b), respectively, for different levels of scintillation index. It should be noted that the SI includes all the parameters such as laser wavelength and beam divergence, aperture averaging, the refractive-index structure parameter, link distance, transmitting and receiving aperture diameters, among others. It is readily seen, from Fig. 14, the tremendous impact of atmospheric turbulence in wireless optical communication systems regardless the modulation scheme being used, although, a PPM system in general performs better than OOK respect to the average optical power needed for achieving a desired bit-error rate level. For example, both for OOK and 8-PPM, there is roughly a 7dB power penalty for a BER = 10^{-6} with fixed threshold receiver (solid lines) and $\sigma_I^2 = 0.5$ respect to the case of no turbulence, i.e. free-space conditions.

For a more comprehensive analysis of a pulse position modulation receiver Fig. 15 is presented. The performance of PPM maximum-likelihood receiver with modulation orders up to 16 is shown in Fig. 15(a), where it becomes evident that increasing the PPM modulation order to the next permitted one there is an improvement of 3dB respect to the average optical power needed to achieve a certain BER. Figure 15(b) is presented as a mean of comparison between the performance of a maximum-likelihood receiver and an optimum threshold receiver for PPM modulation. The penalty incurred for using the optimum threshold receiver instead of the maximum-likelihood receiver is about 1dB for BER < 10^{-4}. Nevertheless, this penalty increases for higher values of bit-error rate, although, a typical communication system with no forward error correction (FEC) code implemented will be designed to have a BER lower than 10^{-4}.

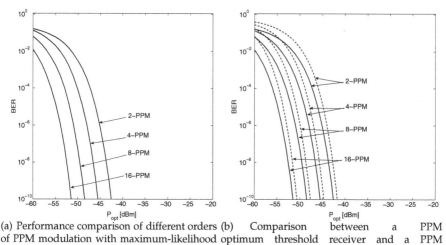

(a) Performance comparison of different orders (b) Comparison between a PPM of PPM modulation with maximum-likelihood optimum threshold receiver and a PPM receiver. maximum-likelihood receiver (solid line).

Fig. 15. Performance of a PPM optimal threshold receiver and PPM maximum-likelihood receiver vs received average optical power under free-space (no-turbulence) conditions.

For the calculations in Fig. 14 and Fig. 15 background radiation was omitted in the analysis and it was assumed an APD's current gain $G = 70$, $k_{eff} = 0.094$, quantum efficiency $\eta = 0.77$ and $NEP = 55\text{fW}/\sqrt{\text{Hz}}$. The light source was set to $\lambda = 780\text{nm}$, pulse time $T_p = 1\text{ns}$ and extinction ratio $\epsilon = 0.001$.

7. References

Al-Habash, M. A., Andrews, L. C. & Philips, R. L. (2001). Mathematical model for the irradiance probability density function of a laser beam propagating through turbulent media, *Opt. Eng.* 40(8): 1554–1562.

Andreas, E. L. (1988). Estimating c_n^2 over snow and sea ice from meteorological data, *J. Opt. Soc. Am. A* 5(4): 481–495.

Andrews, L. C. (1992). Aperture-averaging factor for optical scintillations of plane and spherical waves in the atmosphere, *J. Opt. Soc. Am.* 9(4): 597–600.

Andrews, L. C., Al-Habash, M. A., Hopen, C. Y. & Phillips, R. L. (2001). Theory of optical scintillation: Gaussian-beam wave model, *Waves in Random and Complex Media* 11(3): 271–291.

Andrews, L. C. & Philips, R. L. (2005). *Laser Beam Propagation through Random Media*, 2nd edn, SPIE Press, Belligham.

Arimoto, Y. (2010). Near field laser transmission with bidirectional beacon tracking for tbps class wireless communications, *in* H. Hemmati (ed.), *Free-Space Laser Communication Technologies XXII*, Vol. 7587 of *Proc. SPIE*, pp. 758708-1–758708-8.

Barry, J. (1994). *Wirelees Infrarred Communications*, The Kluwer International Series in Engineering and Computer Science, Kluwer Academic Publishers, Norwell.

Bird, R. E. & Riordan, C. (1986). Simple solar spectral model for direct and diffuse irradiance on horizontal and tilted plans at the earth's surface for cloudless atmosphere, *J. Clim. Meteor.* 25: 87–97.

Cazaubiel, V., Troy, B. & Chene, B. (2006). Optical DATA relay satellite architecture and airborne laser communication demonstrator, *Emerging and Future Technologies for Space Based Operations Support to NATO Military Operations*, Proc. RTO-MP-RTB-SPSM-001, pp. 20-1–20-10.

Chan, V. (1982). Coding for the turbulent atmospheric optical channel, *IEEE Trans. Commun.* 30(1): 269–275.

Churnside, J. H. (1991). Aperture averaging of optical scintillations in the turbulent atmosphere, *Appl. Opt.* 30: 1982–1994.

Churnside, J. H. & Frehlich, R. G. (1989). Experimental evaluation of log-normally modulated rician and ik models of optical scintillation in the atmosphere, *J. Opt. Soc. Am. A* 6(11): 1760–1766.

Churnside, J. H. & Lataitis, R. J. (1990). Wander of an optical beam in the turbulent atmosphere, *Appl. Opt.* 29(7): 926–930.

Ciaramella, E., Arimoto, Y., Contestabile, G., Presi, M., D'Errico, A., Guarino, E. & Matsumoto, M. (2009). 1.28 Terabit/s (32x40 Gbit/s) WDM transmission system for free space optical communications, *IEEE J. Sel. Area Comm.* 27: 1639–1645.

Davidson, F. M. & Sun, X. (1988). Gaussian approximation versus nearly exact performance analysis of optical communication systems with ppm signaling and apd receivers, *IEEE. Trans. Commun.* 36(11): 1185–1192.

Dios, F., Recolons, J., Rodríguez, A. & Batet, O. (2008). Temporal analysis of laser beam propagation in the atmosphere using computer-generated long phase screens, *Opt. Express* 16(3): 2206–2220.

Dolinar, S. J., Hamkins, J., Moision, B. E. & Vilnrotter, V. A. (2006). *Deep Space Optical Communications*, Wiley-Interscience, chapter Optical Modulation and Coding, pp. 215–299.

Doss-Hammel, S., Oh, E., Ricklin, J., Eaton, F., Gilbreath, C. & Tsintikidis, D. (2004). A comparison of optical turbulence models, *in* J. C. Ricklin & D. G. Voelz (eds), *Remote Sensing and Modeling of Ecosystems for Sustainability*, Vol. 5550 of *Proc. SPIE*, pp. 236–246.

Epple, B. (2010). Simplified channel model for simulation of free-space optical communications, *J. Opt. Commun. Netw.* 2(5): 293–304.

Fante, R. L. (1980). Electromagnetic beam propagation in turbulent media: an update, *Proc. IEEE* 68(11): 1424–1443.

Fleck, J. A., Morris, J. R. & Feit, M. D. (1976). Time-dependent propagation of high-energy laser beams through the atmosphere, *Appl. Phys.* 10: 129–160.

Fletcher, G., Hicks, T. & Laurent, B. (1991). The SILEX optical interorbit link experiment, *J. Electron. Comm. Eng.* 3(6): 273–279.

Frehlich, R. (2000). Simulation of laser propagation in a turbulent atmosphere, *J. Opt. Soc. Am.* 39(3): 393–397.

Fried, D. L. (1967). Aperture averaging of scintillation, *J. Opt. Soc. Am.* 57(2): 169–175.

Georghiades, C. N. & Snyder, D. L. (1984). Locating data frames in direct-detection optical communication systems, *IEEE Trans. Commun.* COM-32(2): 118–123.

Harding, C. M., Johnston, R. A. & Lane, R. G. (1999). Fast simulation of a kolmogorov phase screen, *Appl. Opt.* 38(11): 2161–2160.

Hudson, R. D. (1969). *Infrared System Engineering*, John Wiley & Sons, New York.

Ishimaru, A. (1969). Fluctuations of a focused beam wave for atmospheric turbulence probing, *Proc. IEEE* 57: 407–414.

Jeganathan, M., Toyoshima, M., Wilson, K. ., James, J., Xu, G. & Lesh, J. (1997). Data analysis results from the GOLD experiments, *Free-Space Laser Comm. Tech. IX*, Vol. 1990 of *Proc. SPIE*, pp. 70–81.

Kim, I. I., Stieger, R., Koontz, J. A., Moursund, C., Barclay, M., Adhikari, M., Schuster, J., Korevaar, E., Ruigrok, R. & DeCusatis, C. (1998). Wireless optical transmission of fast ethernet, FDDI, ATM, and ESCON protocol data using the TerraLink laser communication system, *Opt. Eng.* 37(12): 3143–3155.

Knestrick, G. L. & Curcio, J. A. (1967). Measurements of spectral radiance of the horizon sky, *Appl. Opt.* 6(12): 2105–2109.

Kolmogorov, A. N. (1941). The local structure of turbulence in a incompressible viscous fluid for very large reynolds numbers, *C. R. (Doki) Acad. Sci. U.S.R.R.* 30: 301–305.

Lane, R. G., Glindemann, A. & Dainty, J. C. (1992). Simulation of a kolmogorov phase screen, *Waves Random Media* 2: 209–224.

Lawrence, R. S., Ochs, G. R. & Clifford, S. F. (1970). Measurements of atmospheric turbulence relevant to optical propagation, *J. Opt. Soc. Am.* 60(6): 826–830.

Levine, B., Martinsen, E., Wirth, A., Jankevics, A., Toledo-Quinones, M., Landers, F. & Bruno, T. (1998). Horizontal line-of-sight turbulence over near-ground paths and implications for adaptive optics corrections in laser communications, *Appl. Opt.* 37(21): 4782–4788.

Martin, J. M. & Flattè, S. M. (1998). Intensity images and statistics from numerical simulation of wave propagation in 3-d random media, *J. Opt. Soc. Am.* 27(11): 2111–2126.

Moreira, A. J. C., Valdas, R. T. & de Oliveira Duarte, A. M. (1996). Performance of infrared transmission systems under ambient light interference, *IEE Proc. Optoelectron.* 143(6): 339–346.

Ochs, G. R. & Hill, R. J. (1985). Optical-scintillation method of measuring turbulence inner scale, *Appl. Opt.* 24(15): 2430–2432.

Okazaki, A. (1978). Pulse interval modulation applicable to narrow band transmission, *IEEE. Trans. on Cable Television* CATV-3(4): 155–164.

Otte, R., de Jong, L. P. & van Roermund, A. H. M. (1998). Slot synchronization by reducing the ppm pulsewidth in wireless optical systems, *IEEE Trans. on Circuits and Systems* 45(7): 901–903.

Patarasen, S. & Georghlades, C. N. (1992). Frame synchronization for optical overlapping pulse-position modulation systems, *IEEE Trans. Commun.* 40(4): 783–794.

Perez-Jimenez, R., Rabadan, J., Melian, V. & Betancor, M. (1996). Improved ppm modulations for high spectral efficiency ir-wlan, *Personal, Indoor and Mobile Radio Communications, 1996. PIMRC'96., Seventh IEEE International Symposium on*, Vol. 1, pp. 262–266.

Perlot, N. & Fritzsche, D. (2004). Aperture-averaging - theory and measurements, *in* G. S. Mecherle, C. Y. Young & J. S. Stryjewski (eds), *Free-Space Laser Communication Technologies XVI*, Vol. 5338 of *Proc. SPIE*, pp. 233–242.

Proakis, J. G. (2001). *Digital Communications*, McGraw-Hill, Boston.

Recolons, J., Andrews, L. & Philips, R. L. (2007). Analysis of beam wander effects for a horizontal-path propagating gaussian-beam wave: focused beam case, *Opt. Eng.* 46(8): 086002-1–086002-11.

Recolons, J. & Dios, F. (2005). Accurate calculation of phase screens for the modelling of laser beam propagation through atmospheric turbulence, in S. M. Doss-Hammel & A. Kohnle (eds), *Atmospheric Optical Modeling, Measurement, and Simulation*, Vol. 5891 of *Proc. SPIE*, pp. 51–62.

Ricklin, J. C., Bucaille, S. & Davidson, F. M. (2004). Performance loss factors for optical communication through clear air turbulence, in D. G. Voelz & J. C. Ricklin (eds), *Free-Space Laser Communications and Active Laser Illumination III*, Vol. 5160 of *Proc. SPIE*, pp. 1–12.

Ricklin, J. C., Hammel, S. M., Eaton, F. D. & Lachinova, S. L. (2006). Atmospheric channel effects on free-space laser communication, *J. Opt. Fiber. Commun. Rep.* 3: 111–158.

Sadot, D. & Kopeika, N. S. (1992). Forecasting optical turbulence strength on the basis of macroscale meteorology and aerosols: models and validation, *Opt. Eng.* 31(31): 200–212.

Shiu, D. & Kahn, J. M. (1999). Differential pulse-position modulation for power-efficient optical communication, *IEEE Trans. on Commun.* 47(8): 1201–1211.

Sigiyama, H. & Nosu, K. (1989). MPPM: A method for improving the band-utilization efficiency in optical PPM, *IEEE J. Lightwave Techn.* 7(3): 465–472.

Sova, R. M., Sluz, J. E., Young, D. W., Juarez, J. C., Dwivedi, A., Demidovich, I. N. M., Graves, J. E., Northcott, M., Douglass, J., Phillips, J., Driver, D., McClarin, A. & Abelson, D. (2006). 80 gb/s free-space optical communication demonstration between an aerostat and a ground station, in A. K. Majumdar & C. C. Davis (eds), *Free-Space Laser Communications VI*, Vol. 6304 of *Proc. SPIE*, p. 630414.

Srinivasan, M., Vilnrotter, V. & Lee, C. (2005). Decision-directed slot synchronization for pulse-position-modulated optical signals, *The Interplanetary Network Progress Report 42-161*, Jet Propulsion Laboratory, Pasadena, California.

Sun, X. & Davidson, F. M. (1990). Word timing recovery in direct detection optical ppm communication systems with avalanche photodiodes using a phase lock loop, *IEEE Trans. Commun.* 38(5): 666–673.

Tatarskii, V. I. (1971). *The effects of the turbulent atmosphere on wave propagation*, Israel Program for Scientific Translation, Jerusalem.

Tyson, R. K. (2002). Bit-error rate for free-space adaptive optics laser communications, *J. Opt. Soc. Am. A* 19(4): 753–758.

Vetelino, F. S., Young, C. & Andrews, L. (2007). Fade statistics and aperture averaging for gaussian beam waves in moderate-to-strong turbulence, *Appl. Opt.* 46(18): 3780–3790.

Wang, T., Ochs, G. R. & Clifford, S. F. (1978). A saturation-resistant optical scintillometer to measure C_n^2, *J. Opt. Soc. Am.* 68(3): 334–338.

Wayne, D. T., Philips, R. L. & Andrews, L. C. (2010). Comparing the log-normal and gamma-gamma model to experimental probability density functions of aperture averaging data, in C. Majumdar, A. K.and Davis (ed.), *Free-Space Laser Communications X*, Vol. 7814 of *Proc. SPIE*, p. 78140K.

Webb, P. P., McIntyre, R. J. & Conradi, J. (1974). Properties of avalanche photodiodes, *RCA review* 35: 234–278.

Weyrauch, T. & Vorontsov, M. A. (2004). Free-space laser communications with adaptive optics: Atmospheric compensation experiments, *J. Opt. Fiber. Commun. Rep.* 1: 355–379.

Wong, K. K., R, T. O. & Kiatweerasakul, M. (2000). The performance of optical wireless ook, 2-ppm and spread spectrum under the effects of multipath dispersion and artificial light interference, *Int. J. Commun. Syst.* 13: 551–576.

Yitzhaky, Y., Dror, I. & Kopeika, N. S. (1997). Restoration of atmospherically blurred images according to weather-predicted atmospheric modulation transfer functions, *Opt. Eng.* 36(11): 3064–3072.

Yura, H. T., Sung, C. C., Clifford, S. F. & Hill, R. J. (1983). Second-order rytov approximation, *J. Opt. Soc. Am.* 73(4): 500–502.

2

Full-Field Detection with Electronic Signal Processing

Jian Zhao and Andrew D. Ellis
Photonic Systems Group, Tyndall National Institute & Department of Physics,
University College Cork
Ireland

1. Introduction

The rapid growth in broadband services is increasing the demand for high-speed optical communication systems. However, as the data rate increases, transmission impairments such as chromatic dispersion (CD) become prominent and require careful compensation. In addition, it is proposed that the next-generation optical networks will be intelligent and adaptive with impairment compensation that can be software-defined and re-programmed to adapt to changes in network conditions. This flexibility should allow dynamic resource reallocation, provide greater network efficiency, and reduce the operation and maintenance cost. Conventional dispersion compensating fiber (DCF) is bulky and requires careful design for each fiber link as well as associated amplifiers and monitoring. Recently, the advance of high-speed microelectronics, for example 30 GSamples/s analogue to digital converters (ADC) (Ellermeyer et al., 2008), has enabled the applications of electronic dispersion compensation (EDC) (Iwashita & Takachio, 1988; Winters & Gitlin, 1990) in optical communication systems at 10 Gbaud and beyond. The maturity in electronic buffering, computation, and large scale integration enables EDC to be more cost-effective, adaptive, and easier to integrate into transmitters or receivers for extending the reach of legacy multimode optical fiber links (Weem et al., 2005; Schube & Mazzini, 2007) as well as metro and long-haul optical transmission systems (Bülow & Thielecke, 2001; Haunstein & Urbansky, 2004; Xia & Rosenkranz, 2006; Bosco & Poggiolini, 2006; Chandrasekhar et al., 2006; Zhao & Chen, 2007; Bulow et al., 2008). Transmitter-side EDC (McNicol et al., 2005; McGhan et al., 2005 & 2006) exhibits high performance but its adaptation speed is limited by the round-trip delay. Receiver-side EDC can adapt quickly to changes in link conditions and is of particular value for future transparent optical networks where the reconfiguration of the add- and drop-nodes will cause the transmission paths to vary frequently. Direct-detection maximum likelihood sequence estimation (DD MLSE) receivers are commercially available and have been demonstrated in various transmission experiments (Farbert et al., 2004; Gene et al., 2007; Alfiad et al., 2008). However, the performance of conventional EDC using direct detection (DD) is limited due to the loss of the signal phase information (Franceschini et al., 2007). In addition, the transformation of linear optical impairments arising from CD into nonlinear impairments after square-law detection significantly increases the operational complexity of the DD EDC. For example, DD MLSE was numerically predicted to achieve 700km single mode fiber (SMF) transmission at 10 Gbit/s but required 8192 Viterbi processor states (Bosco & Poggiolini, 2006).

This performance limitation could be removed by restoring channel linearity using coherent detection (Taylor, 2004; Tsukamoto et al., 2006; Savory et al., 2007; Cai, 2008) or non-coherent detection based full-field reconstruction (Ellis & McCarthy, 2006; Kikuchi et al., 2006; Liu & Wei, 2007; Polley & Ralph, 2007; Liu et al., 2008; Zhao et al., 2008; Kikuchi & Sasaki, 2010). Digital signal processing (DSP) based coherent detection has various advantages, such as optimized receiver sensitivity and near-ideal impairment compensation capability, facilitating high-speed optical communication systems to approach the capacity limits (Ellis et al., 2010). However, it is expensive, requiring a narrow-linewidth laser, two 90^0 hybrids, four balanced photodiodes, and four ADCs, so it may not be suitable for cost-sensitive applications such as 10-40 Gbit/s access/metro networks and Ethernet. In contrast, the full-field detection (FFD) based systems, which extract the optical field using non-coherent optical receiver and electronic field reconstruction, may greatly relax the complexity. This non-coherent receiver consists of one (or two) asymmetric Mach-Zehnder interferometer (AMZI) and one (or two) pair of matched photodiodes. The recovered optical field then allows for electronic signal processing techniques, including feed-forward equalization (FFE) (Polley & Ralph, 2007; Zhao et al., 2010) and MLSE (McCarthy et al., 2008; Zhao et al., 2009), to improve the performance. When compared to conventional DD systems, FFD linearly transforms the optical signal into the electrical domain, avoiding the nonlinear processing of an impaired signal which constrains the performance of EDC. Consequently, cost-effective frequency-domain equalization is enabled and has been demonstrated to support 10 Gbit/s on-off keying (OOK) transmission over 500 km BT Ireland's field-installed single-mode fiber (SMF) (McCarthy et al., 2009, Zhao et al., 2009), and 50% performance improvement has been achieved for FFD MLSE when compared to DD MLSE using the same number of states (Zhao et al., 2010). Adaptive transmission experiment for a distance range of 0-900 km with less than 400 ns adaptation time was also demonstrated (Zhao & Ellis, 2011). In this chapter, after an analytical discussion of the basic principles, full-field detection based systems are investigated through numerical simulations and experimental demonstration. The essential design criteria to optimize the FFD system performance are discussed and different FFD EDC techniques (frequency-domain equalization, FFE, and MLSE) are examined in terms of performance, complexity, adaptation speed, and robustness to non-optimized parameters. Adaptive transmission with less than 400 ns adaptation time for a wide range of distances from 0 to 900 km is reported using hybrid FFD EDC combining the static frequency-domain equalization and parametric channel estimation based FFD MLSE. The emphasis of this chapter is placed on the amplitude-modulated OOK format, but the investigation on an offset differential quadrature phase shifted keying (offset DQPSK) system is also included in Section 6.

2. Full-field detection

2.1 Basic principle

Fig. 1 depicts the basic principle of full-field detection. The incoming optical signal is processed by a single AMZI with a differential time delay (DTD) of Δt. Assuming that the input optical field (baseband representation) is $E(t)$ $(=\mid E(t)\mid \cdot \exp(j\varphi(t)))$, where $\varphi(t)$ is the phase of the optical field, the two outputs of the AMZI, $E_1(t)$ and $E_2(t)$, are given by:

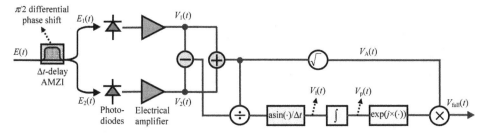

Fig. 1. Basic principle of full optical-field detection. Black and brown lines represent optical and electrical signals respectively. AMZI: asymmetric Mach-Zehnder interferometer.

$$E_1(t) = (E(t) + E(t - \Delta t)e^{j\varphi_{AMZI}}) / 2 \tag{1.1}$$

$$E_2(t) = (E(t) - E(t - \Delta t)e^{j\varphi_{AMZI}}) / 2 \tag{1.2}$$

where φ_{AMZI} is the differential phase of the AMZI. $E_1(t)$ and $E_2(t)$ are detected by a pair of photodiodes and electrically amplified to obtain the electrical signals $V_1(t)$ and $V_2(t)$:

$$V_1(t) \propto [|E(t)|^2 + |E(t - \Delta t)|^2 + 2|E(t)| \cdot |E(t - \Delta t)| \cdot \cos(\varphi(t) - \varphi(t - \Delta t) - \varphi_{AMZI})] / 4 \tag{2.1}$$

$$V_2(t) \propto [|E(t)|^2 + |E(t - \Delta t)|^2 - 2|E(t)| \cdot |E(t - \Delta t)| \cdot \cos(\varphi(t) - \varphi(t - \Delta t) - \varphi_{AMZI})] / 4 \tag{2.2}$$

If we choose $\varphi_{AMZI}=\pi/2$ and small Δt value such that $| E(t-\Delta t) \approx E(t) |$, signals proportional to the intensity, instantaneous frequency, and phase of the optical field, $V_A(t)$, $V_f(t)$, and $V_p(t)$, can be extracted by signal processing of $V_1(t)$ and $V_2(t)$:

$$V_A(t) = [V_1(t) + V_2(t)]^{1/2} = [(|E(t)|^2 + |E(t - \Delta t)|^2) / 2]^{1/2} \approx |E(t - \Delta t / 2)| \tag{3.1}$$

$$V_f(t) = \text{asin}([V_1(t) - V_2(t)] / [V_1(t) + V_2(t)]) / \Delta t \approx (\varphi(t) - \varphi(t - \Delta t)) / \Delta t \tag{3.2}$$

$$V_p(t) = \int V_f(\tau)d\tau = \varphi(t) \tag{3.3}$$

In practice, asin($\cdot$) in Fig.1 can be neglected given $\varphi(t)-\varphi(t-\Delta t)$<<1. By recovering the optical intensity and phase, the full optical field can be reconstructed by:

$$V_{full}(t) \approx V_A(t + \Delta t / 2) \cdot \exp(j \cdot V_p(t)) \tag{4}$$

This full-field reconstruction module may be implemented using analogue (Ellis & McCarthy, 2006) or digital (McCarthy et al., 2009) devices. In the latter case, ADCs are used to sample and quantize $V_1(t)$ and $V_2(t)$.

The configuration in Fig. 1 can be alternatively implemented by Fig. 2, where a pair of balanced photodiodes gives V_1-V_2 and an additional single photodiode obtains V_1+V_2. In the following discussions, we define two new quantities: $V_-(t)=V_1(t)-V_2(t)$ and $V_+(t)=V_1(t)+V_2(t)$.

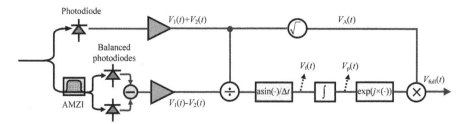

Fig. 2. An alternative implementation of full-field detection

2.2 Optimization of $V_f(t)$ estimation

Fig. 1&2 depict the basic principle for ideal FFD. In practice, additional components are required in the scheme to overcome the susceptibility to noise and associated noise amplification mechanisms during the full-field reconstruction process. The degradation mechanisms discussed in this subsection are induced in the $V_f(t)$ estimation and can be understood from Eq. (3.2), where it is shown that $\varphi(t)-\varphi(t-\Delta t)$ is obtained by the division by the instantaneous received signal power ($V_+(t)$). It is apparent that for the sampling points where the received power is low, this estimation is particularly sensitive to noise. In an OOK-based system where logic data '0' is represented by zero or low power level, this mechanism results in pattern effect with performance degradation for a sequence of consecutive logic data '0's (Zhao et al., 2009). This may be ameliorated by reducing the transmitter extinction ratio (ER) for an OOK system at the expense of a small ER-induced penalty, so the ER should be optimized to maximize the system performance (Zhao et al., 2008). This mechanism also applies to other modulation formats. For example, in a phase-shifted keying system using Mach-Zehnder modulator to encode the phase information, the near-zero intensity due to π phase shift between symbols needs to be controlled, especially in the presence of CD.

Practical pre-amplified optical communication systems usually have an optical power level incident on the photodiodes sufficiently larger than the thermal noise level to ensure that the system is primarily limited by optical amplified spontaneous emission (ASE) noise. However, in FFD, the thermal noise of photodiodes may play an important role in the performance due to the aforementioned low intensity regions. The impact of thermal noise therefore needs to be considered in the overall design of the full-field detector front end. Considering thermal noise of photodiodes, Eq. (3.2) can be re-written as:

$$V_f(t) \propto \operatorname{asin}\left(\frac{|E(t-\Delta t/2)|^2 \sin(\varphi(t)-\varphi(t-\Delta t))+n_{th\_1}-n_{th\_2}}{|E(t-\Delta t/2)|^2+n_{th\_1}+n_{th\_2}}\right)/\Delta t \tag{5}$$

where n_{th_1} and n_{th_2} represent the thermal noise on $V_1(t)$ and $V_2(t)$ in Fig. 1 respectively. It can be seen from Eq. (5) that even when $<|E(t-\Delta t/2)|^2>$, where $<\cdot>$ is the ensemble average, is sufficiently larger than the thermal noise level, the signal-independent thermal noise may have significant impact on the performance for a logical data '0' or the case that the phase difference $\varphi(t)-\varphi(t-\Delta t)$ is small. As discussed above, the use of a lower ER can alleviate this problem (also that caused by the ASE noise) at the expense of back-to-back receiver sensitivity. In practice, a DC bias can be added to $V_+(t)$ before division to mitigate this thermal noise induced effect without the sacrifice of ER. This DC bias may increase the value of the denominator in Eq. (5), which for the signal sequence with low optical intensity, would significantly reduce the impact of thermal noise albeit at the expense of a slight distortion in the reconstructed frequency. Note that in most practical systems, a DC bias would be required in any case in order to accommodate the AC coupling of the receiver. Another effect arising from the thermal noise is attributed to the numerator in Eq. (5) which is approximately linearly proportional to the phase difference $\varphi(t)-\varphi(t-\Delta t)$ and consequently dependent on the DTD of the AMZI. By employing an AMZI with a larger DTD, $\varphi(t)-\varphi(t-\Delta t)$ is increased, which therefore improves the signal to thermal noise ratio. Note that Δt cannot be increased indefinitely because the derivation is based on the assumption of $|E(t-\Delta t)| \approx E(t)|$. The DTD of the AMZI should be designed to a balance between values favouring precise estimation of $V_f(t)$ and thermal noise. In practice, a DTD value between 20%-50% of the symbol period would obtain optimal performance.

2.3 Optimization of $V_p(t)$ estimation

In addition to the design to ensure the accuracy of the $V_f(t)$ estimation, the performance of the full optical-field reconstruction also depends on the quality of phase estimation using $V_f(t)$, which is found to be degraded by low-frequency amplification (Zhao et al., 2008). To illustrate the origin of such an impairment, we may take the Fourier transform to relate the estimated phase $V_p(t)$ to the estimated frequency $V_f(t)$:

$$V_p(t) = \int_0^t V_f(\tau)d\tau \quad \xrightarrow{F} \quad \psi_p(\omega) = \psi_f(\omega)/(j\omega) \tag{6}$$

where $\psi_p(\omega)$ and $\psi_f(\omega)$ are the spectra of $V_p(t)$ and $V_f(t)$ respectively. It is clear from Eq. (6) that the low-frequency components of $V_f(t)$ dominate phase reconstruction, with a scaling factor of $1/|\omega|$. Therefore, any noise or inaccuracy in the low-frequency components of the estimated $V_f(t)$ will accumulate and eventually limit the performance. This suggests that the low-frequency components of $V_f(t)$ should be minimized, which can be achieved by using a high-pass electrical filter before the integrator in Fig. 1&2. Fig. 3 depicts the full configuration of full-field reconstruction. A DC bias is added to the detected signal intensity $V_+(t)$ to accommodate the AC coupling of the receiver and to enhance the robustness to the thermal noise. The amplitude of the optical field $V_A(t)$ is equal to the square root of the re-biased version of $V_+(t)$. A high-pass filter is placed in the phase estimation path to suppress the low-frequency amplification. The optical phase $V_p(t)$ is obtained by integrating the frequency $V_f(t)$ and employed to reconstruct the full optical field $V_{full}(t)$ using an exponential function and a multiplier.

3. Electronic signal processing techniques for dispersion compensation

The recovered optical field allows for subsequent dispersion compensation using electrical-domain signal processing techniques, as shown in Fig. 3. The full-field approaches offer a compromise between conventional cost-effective DD EDC that lacks phase information and is thus limited in performance, and coherent-detection based EDC (see Fig. 4) that has better performance but is expensive, requiring a narrow linewidth laser, two 90⁰ hybrids, four pairs of balanced photodiodes and four ADCs. Although FFD needs additional full-field reconstruction, it avoids complicated estimation of the frequency offset, polarization and phase difference between the signal and the local oscillator that are required in coherent detection. The complexity of the dispersion compensation module in FFD is also comparable to that in coherent detection.

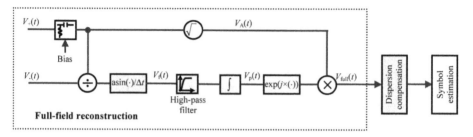

Fig. 3. Full configuration of full-field reconstruction with a bias added to $V_+(t)$ to mitigate the thermal noise effect and a high-pass filter in the phase estimation path to suppress low-frequency amplification.

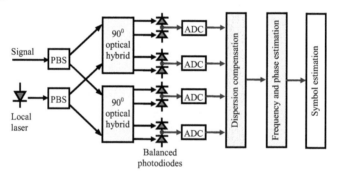

Fig. 4. Configuration of coherent detection. PBS: polarization beam splitter.

3.1 Frequency-domain equalization

By recovering the full optical field, the linearity of the channel is preserved and linear impairments such as CD can be simply compensated by applying their inverse transfer function. For large values of accumulated dispersion, this is optimally applied in the frequency domain. For an analogue signal, CD would be compensated using a dispersive microstripline (McCarthy & Ellis, 2007). In the digital domain (see Fig. 5(a)), this is implemented using frequency-domain equalization where we convert the recovered optical field $V_{full}(t)$ into parallel blocks, take the fast Fourier transform (FFT) for each block, multiply the transformed signal spectrum by the inverse transfer function of CD, take the inverse fast Fourier transform (IFFT) and then convert the blocks into compensated serial time-domain signal. Each block has overlaps in time with its adjacent blocks to allow for the guard interval for CD compensation (see the inset of Fig. 5(a)), whose length should be longer than the memory length of the channel intersymbol interference (ISI). The inverse transfer function of CD is:

$$H(\omega) = \exp(-j\beta_2 z \omega^2 / 2) \tag{7}$$

where $\beta_2 z$ is the accumulated CD value and ω is the frequency. With this technique, only $\beta_2 z$ needs to be controlled, and is expected to match the CD of the actual fiber link.

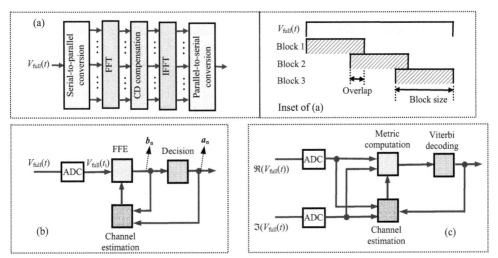

Fig. 5. Configurations of (a) frequency-domain equalization; (b) FFE; (c) MLSE. Inset of (a) shows the block-based processing of frequency-domain equalization, with overlaps between adjacent blocks as the guard interval.

3.2 Full-field detection feed-forward equalization (FFD FFE)

For small CD values where the memory length of the intersymbol interference is not large, the compensation may be implemented using a time-domain filter with an appropriate response. One implementation is a finite impulse response filter comprising a cascade of tapped and weighted delay lines known as feed-forward equalizer (FFE). FFE is suitable for implementation using either analogue (Haunstein & Urbansky, 2004) or digital circuits. For the digital implementation (see Fig. 5(b)), the digital signal representing the recovered field $V_{full}(t)$, $V_{full}(t_i)$, is processed to give the estimated sequence, b_n:

$$b_n = \sum_{i=-Nm/2}^{Nm/2} V_{full}(t_{n-i/N}) \cdot f_{i/N} \tag{8}$$

where $f_{i/N}$, $-Nm/2 \leq i \leq Nm/2$, is the FFE coefficient with N being the sample number per bit and m being the memory length. The tap weights $f_{i/N}$ are updated by comparing the values of the estimates with the values after decision or the training sequence data (Proakis, 2000):

$$f_{i/N}^{(n+1)} = f_{i/N}^{n} + \Delta \cdot (a_n - b_n) \cdot V_{full}(t_{n-i/N}) \tag{9}$$

where Δ is a parameter to control the update speed. a_n is the nth decoded data and is replaced by the training sequence during initial channel estimation.

3.3 Full-field detection maximum likelihood sequence estimation (FFD MLSE)

In MLSE, rather than compensating for the CD-induced distortion prior to symbol decision, the DSP circuit builds up a "channel model" representing the expected received waveforms for a complete set of transmitted sequences. These stored waveforms are then compared to the actual received waveform and the sequence that most likely results in the received waveform is selected. In practice, the "channel model" can be simplified by assuming a finite channel memory length and the comparison process may be performed using a recursive algorithm proposed by Viterbi (Proakis, 2000). In FFD MLSE (see Fig. 5(c)), the real and imaginary information are both exploited for building up the "channel model" (formally known as channel training) and for calculating the probability that the received waveform matches one of the stored waveforms (formally known as metric computation). Mathematically, the metric of FFD MLSE, $PM(a_n)$, is calculated as (McCarthy et al., 2008):

$$PM(a_n) = PM(a_{n-1}) - \sum_i \log(p(\Re(V_{full}(t_i)), \Im(V_{full}(t_i)) \mid a_{n-m}, ..., a_n)) \qquad (10)$$

where i represents the samples associated with the bit n ($i \in \{n, n+1/2\}$ for two samples per bit). $\Re(\cdot)$ and $\Im(\cdot)$ represent the real and imaginary components. a_n and $p(\Re(V_{full}(t_i)), \Im(V_{full}(t_i)) \mid a_{n-m}, ..., a_n)$ are the nth logical data and the two-dimensional joint probability of the full optical field at time t_i given the logical data $a_{n-m}, ..., a_n$, respectively. m is the memory length. The initial joint probabilities are obtained using either a histogram or a parametric method. In the histogram method, lookup table is established for $p(\Re(V_{full}(t_i)), \Im(V_{full}(t_i)) \mid a_{n-m}, ..., a_n)$ with the table size proportional to 2^{2q+m+2} at a sampling rate of two samples per bit, where q is the ADC resolution. The complexities of the metric computation and the Viterbi decoding are the same as those of a DD MLSE with the same state number, and are proportional to 2^{m+1} and 2^m respectively. In practice, the full expression of metric (10) can be approximated by assuming that the probability distributions for the real and imaginary signals are independent, giving a new metric (Zhao et al., 2009):

$$PM(a_n) = PM(a_{n-1}) - \sum_i \log(p(\Re(V_{full}(t_i)) \mid a_{n-m}, ..., a_n) \cdot p(\Im(V_{full}(t_i)) \mid a_{n-m}, ..., a_n)) \qquad (11)$$

This simplification causes only a slight performance penalty when used with optimized system parameters, but significantly reduces the required lookup table size and the time for lookup table setup and update from 2^{2q+m+2} to 2^{q+m+3}.

Lookup table based histogram channel estimation is precise, and, to a certain extent, able to mitigate nonlinear impairments which distort the signal in a deterministic manner. However, the required training sequence to obtain the lookup table may be long. On the other hand, parametric channel estimation, where the lookup table is obtained based on the assumption of a distribution for the received samples and the calculation of a few basic parameters for the distribution, can greatly improve the adaptation speed. By recovering the full optical field, $p(\Re(V_{full}(t_i)) \mid a_{n-m}, ..., a_n)$ and $p(\Im(V_{full}(t_i)) \mid a_{n-m}, ..., a_n)$ can be approximated using Gaussian distribution (Zhao et al., 2011):

$$PM(a_n) = PM(a_{n-1})$$

$$+ \sum_i [\log(\sigma_{i,r}(a_{n-m},...,a_n)) + (\Re(V_{full}(t_i)) - \mu_{i,r}(a_{n-m},...,a_n))^2 / 2 / \sigma_{i,r}(a_{n-m},...,a_n)^2 \quad (12)$$

$$+ \log(\sigma_{i,q}(a_{n-m},...,a_n)) + (\Im(V_{full}(t_i)) - \mu_{i,q}(a_{n-m},...,a_n))^2 / 2 / \sigma_{i,q}(a_{n-m},...,a_n)^2]$$

where $\mu_{i,r}$ and $\mu_{i,q}$ are the means of the real and imaginary tributaries of the signal while $\sigma_{i,r}$, $\sigma_{i,q}$ are the variances, all dependent on the logical data $a_{n-m},...,a_n$.

4. Numerical analysis for 10 Gbit/s FFD-based OOK systems

Fig. 6 shows the simulation model implemented using Matlab. Continuous wave light was intensity modulated by a 10 Gbit/s OOK data train using a Mach-Zehnder modulator (MZM). The data train consisted of a 2^{11}-1 pseudo-random binary sequence (PRBS) repeated nine times (18,423 bits). 10 '0' bits and 11 '0' bits were added before and after this data train respectively to simplify the boundary conditions. The bits were raised-cosine shaped with a roll-off coefficient of 0.4 and had 40 samples per bit. The extinction ratio (ER) of the modulated OOK signal was set by adjusting the bias and the amplitude of the electrical OOK data. The signal was launched into the transmission link with 80 km SMF per span and -3 dBm signal power. The SMF had CD of 16 ps/km/nm, a nonlinear coefficient of 1.2/km/W, and a loss of 0.2 dB/km. The split-step Fourier method was used to calculate the signal propagation in the fibers. At the end of each span, noise from Erbium-doped fiber amplifiers (EDFA) was modelled as complex additive white Gaussian noise with zero mean and a power spectral density of $n_{sp}h\nu(G-1)$ for each polarization, where G and $h\nu$ are the amplifier gain and the photon energy respectively. n_{sp} is population inversion factor of the amplifiers and was set to give 4 dB amplifier noise figure (NF). The noise of the optical preamplifier was also modelled as additive white Gaussian noise with random polarization. The launch power into the preamplifier was adjusted by a variable optical attenuator (VOA) to control the optical signal-to-noise ratio (OSNR). The pre-amplified signal was filtered by an 8.5 GHz Gaussian-shaped optical band-pass filter (OBPF), unless otherwise stated. The signal after the OBPF was then split into two paths to extract $V_-(t)$ and $V_+(t)$. The AMZI for the extraction of $V_-(t)$ had $\pi/2$ differential phase shift and differential time delay (DTD) of either 10 ps or 30 ps. The responsivities of the balanced photodiodes and the direct photodiode were assumed to be 0.6 A/W and 0.9 A/W respectively, and equivalent thermal noise spectral power densities were assumed to be 100 pA/Hz$^{1/2}$ and 18 pA/Hz$^{1/2}$ respectively. These parameters match typical values of commercially available detectors. The optical power incident on the photodiodes was 0 dBm. After detection, the signals were electrically amplified, filtered by 15 GHz 4th-order Bessel electrical filters (EFs), and down-sampled to 50 GSamples/s to simulate the sampling effect of the real-time oscilloscope. $V_+(t)$ was re-biased to allow for the AC coupling of the receiver and to enhance the robustness to thermal noise. The high-pass EF to suppress the low-frequency amplification was Gaussian-shaped. $V_-(t)$ and $V_+(t)$ were exploited to reconstruct the optical signal, which was subsequently compensated using frequency-domain equalization, FFD FFE and FFD MLSE. The simulation was iterated seven times with different random number seeds to give a total of 128,961 simulated bits. The performance was evaluated in terms of the required OSNR (0.1 nm resolution) to achieve a bit error rate (BER) of 5×10^{-4} by direct error counting. 128,961 bits were sufficient to produce a confidence interval of [3.5×10^{-4} 7×10^{-4}] for this BER with 99% certainty (Jeruchim, 1984).

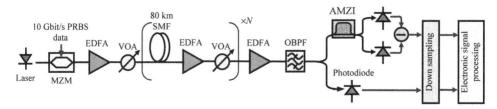

Fig. 6. Numerical model of FFD-based OOK systems. MZM: Mach-Zehnder modulator; EDFA: Erbium-doped fiber amplifier; VOA: variable optical attenuator; OBPF: optical band-pass filter.

4.1 System design based on frequency-domain equalization

In this subsection, simulations are performed to verify the important design rules as described in Section 2. The results are based on frequency-domain equalization, but the developed guidelines also apply to FFD FFE and FFD MLSE.

4.1.1 Optimization of optical-field reconstruction

As discussed in Section 2, it is essential to optimize the system to ensure the quality of optical-field reconstruction. Fig. 7 shows the eye diagrams of the signal at a fiber length of 2160 km using frequency-domain equalization. In these figures, thermal noise and fiber nonlinearity were not included. For a larger ER (Fig. 7(a)), the received value of $V_+(t)$ for a sequence of consecutive logical data '0's was so small that any optical noise led to large estimation inaccuracy in $V_f(t)$. This inaccuracy contained significant low-frequency content, which was further increased by the low-frequency amplification mechanism. By using a smaller ER, the value of $V_+(t)$ for a sequence of consecutive logical data '0's was increased, reducing the estimation inaccuracy of $V_f(t)$ and resulting in better compensation performance, as shown in Fig. 7(b). The high-pass EF in the phase estimation path further reduced the low-frequency components of $V_f(t)$. As a result, the compensated signal after 2160 km shown in Fig. 7(c) has a significantly clearer eye than those in Fig. 7(a) and (b).

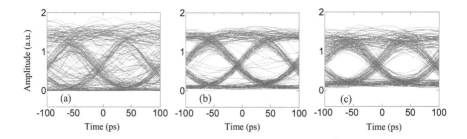

Fig. 7. Eye diagrams of the signal after frequency-domain equalization at 2160 km [(a): 25 dB ER without a high-pass EF; (b): 12 dB ER without a high-pass EF; (c): 12 dB ER with a 0.85 GHz high-pass EF]. Fiber nonlinearity and thermal noise are not included.

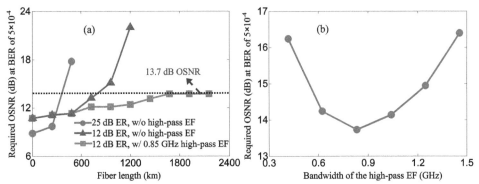

Fig. 8. (a) Required OSNR versus fiber length (circles: 25 dB ER without a high-pass EF; triangles: 12 dB ER without an EF; squares: 12 dB ER with a 0.85 GHz high-pass EF). (b) Required OSNR versus 3 dB bandwidth of the high-pass EF at 2160 km and 12 dB ER. Fiber nonlinear and thermal noise are not included

To quantify the performance improvement of the method, Fig. 8(a) depicts the required OSNR for these three cases. The figure shows that by using 12 dB ER and a 0.85 GHz high-pass EF, the OSNR transmission limit could be significantly extended, despite the back-to-back penalty arising from the reduced ER. At a system length of 2160 km, the required OSNR was around 13.7 dB. It should be noted that whilst the high-pass EF suppressed the impairment from low-frequency amplification, it also introduced distortion to the estimated frequency $V_f(t)$. This distortion resulted in the rails of the eye diagrams in Fig. 7(c) being somewhat thicker than those in Fig. 7(b). Clearly, a trade-off exists between the impairment from low-frequency amplification and the distortion. At an ER of 12 dB and 2160 km, the optimized bandwidth of the EF was around 0.85 GHz as shown in Fig. 8(b).

4.1.2 Impacts of fiber nonlinearity and thermal noise

Fig. 9(a) shows the required OSNR without (circles) and with (triangles) fiber nonlinearity and the maximum achievable OSNR (squares) for 80 km SMF and -3 dBm signal launch power per span. The ER was 12 dB and a 0.85 GHz Gaussian-shaped high-pass EF was employed. The DTD of the AMZI and the bias of $V_+(t)$ were assumed to be 10 ps and 0 V respectively, and photodiode thermal noise was neglected. From the figure, it is shown that including fiber nonlinearity in the transmission simulation resulted in an additional penalty of up to 1.8 dB for system lengths less than 2160 km. On the other hand, the maximum achievable OSNR degraded as the fiber length increased, and the maximum achievable OSNRs were 30.2 dB and 20.6 dB for 240 km and 2160 km respectively. At 2160 km, the maximum achievable OSNR was more than 5 dB greater than the required value. Fig. 9(a) shows the required OSNR when the thermal noise contribution of the receiver was neglected, representing the maximum achievable performance. However, as discussed in Section 2, receiver thermal noise can significantly influence the performance of the FFD EDC schemes. Fig. 9(b) shows the required OSNR as a function of system length without thermal noise (circles) and with thermal noise for various values of key parameters (applied DC bias offset normalized to M, the average detected signal amplitude, and the AMZI DTD). The figure shows that thermal noise may limit the transmission distance to less than 240 km. By employing an AMZI with a larger DTD and biasing the detected intensity signal $V_+(t)$, the tolerance to thermal noise was significantly increased, which was attributed to the improvement in the $V_f(t)$ estimation.

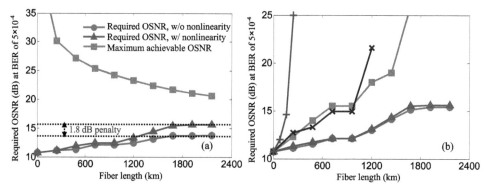

Fig. 9. (a) Required OSNR without (circles) and with (triangles) fiber nonlinearity and the maximum achievable OSNR (squares). Thermal noise is neglected. (b) Required OSNR versus system length without thermal noise under 0 V $V_+(t)$ bias and 10 ps AMZI DTD (circles), and with thermal noise: pluses: 0 V $V_+(t)$ bias and 10 ps AMZI DTD; squares: 0.1M $V_+(t)$ bias and 10 ps AMZI DTD; crosses: 0 V $V_+(t)$ bias and 30 ps AMZI DTD; triangles: 0.1M $V_+(t)$ bias and 30 ps AMZI DTD. Fiber nonlinearity is included.

4.1.3 Impacts of optical band-pass filter bandwidth

The final parameter to be optimized is the OBPF bandwidth. Fig. 10 shows the required OSNR versus the OBPF bandwidth at 6 dB and 12 dB ER for 960 km. $V_+(t)$ bias was 0.1M and the AMZI DTD was 30 ps. A 0.85 GHz high-pass EF was used for suppression of low-frequency amplification. The figure shows that when the ASE noise was not sufficiently suppressed (bandwidth>0.25 nm), a system with 6 dB signal ER exhibited better performance compared to that with 12 dB ER. This matches recent experimental demonstration (McCarthy et al., 2009) and is because a lower ER could effectively reduce noise amplification arising from the division by total received power and the low-frequency amplification in phase estimation. However, if the ASE noise was sufficiently suppressed (bandwidth<0.25 nm), the benefit of reducing the ER was reduced. Clearly, the optimal performance depended on a balance between mitigating the noise amplification and penalty induced by a lower ER. The optimal filter bandwidths for 6 dB and 12 dB ERs were 0.07 nm (~8.5 GHz).

4.2 FFD feed-forward equalizer and comparison to frequency-domain equalization

The design rules developed above are also applicable to FFD FFE and FFD MLSE. In this subsection, the performance of adaptive FFD FFE is numerically investigated and compared to the static frequency-domain equalization. The system parameters are set to the optimal values obtained in Section 4.1, specifically 12 dB ER, 8.5 GHz OBPF bandwidth, 30 ps AMZI DTD, and 0.1M $V_+(t)$ bias. Fiber nonlinearity and thermal noise are included.

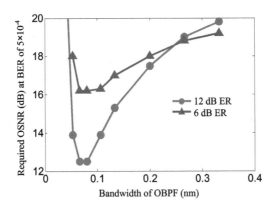

Fig. 10. Required OSNR versus bandwidth of the OBPF at 6 dB ER (triangles) and 12 dB ER (circles) for 960 km. Fiber nonlinearity and thermal noise are included.

Fig. 11(a) shows the performance versus fiber length using frequency-domain equalization manually set to compensate 100% of the accumulated CD for each distance value (circles) and adaptive FFD FFE by initially setting the FFE coefficients to compensate 1080 km CD (triangles). Solid and dashed lines represent the cases using 2 and 5 samples/bit ADCs. In the figure, the performances of FFD FFE for five and two samples per bit were almost the same (<0.2 dB), so only the curve for the case of two samples per bit was plotted. The memory length of FFE was 32 bits (64 taps at 2 samples per bit). It is clearly seen that more than 5 dB OSNR penalty was observed at 2160 km for the frequency-domain equalization when using the reduced sampling rate. This penalty was due to not only the increased calculation inaccuracy during field reconstruction at a lower sampling rate but also the aliasing effect of the ADCs such that the distortion imposed by CD could not be fully compensated in the digital domain by a fixed filter with the exact inverse transfer function of the CD. In contrast, FFD FFE automatically searched the optimal condition to minimize the distortion even with the aliasing effect. Consequently, it exhibited more robustness to the reduction of the sampling rate. Note that due to the capability of compensating ISI regardless of the source, FFD FFE also mitigated other distortions, such as the distortion induced by the high-pass EF used in the phase estimation path. To illustrate this, Fig. 11(b) shows the required OSNR versus the bandwidth of the EF filter at 2160 km for frequency-domain equalization (circles) and adaptive FFD FFE (triangles). The figure shows that the performance of FFE was degraded when the system was dominated by the low-frequency amplification (<0.5 GHz). Consequently, a high-pass EF with sufficient bandwidth was required. However, a sufficiently wide filter bandwidth would result in distortion. When using frequency-domain equalization, the filter bandwidth should be carefully optimized to balance the low-frequency amplification and the distortion. In contrast, FFD FFE was robust to such distortion, resulting in improved performance and wider tolerance range.

In practice, unless precise clock recovery is performed, the common sampling phase of the two paths will drift throughout the eye. The misalignment of the sampling phase, t_0, can be viewed as a filter with transfer function of $\exp(-j\omega t_0)$, where t_0 is unknown and might slowly vary with time. Adaptive filters such as FFE can track and mitigate such distortion. By minimizing the mean square value of the decision error, the coefficients of FFD FFE may be self-adjusted to construct a transfer function equal to the multiplication of the inverse transfer function of CD, sampling phase misalignment, and the remaining ISI effects. Fig. 12(a) shows the simulated performance versus the sampling phase misalignment by using frequency-domain equalization (circles) and FFD FFE (triangles). FFD FFE has 2 samples/bit whilst the frequency-domain equalization employs 5 samples/bit. The figure shows that at 2160 km, FFD FFE exhibited negligible penalty for the sampling phase between [-50 ps 50 ps], which was much more robust when compared to the static frequency-domain equalization (circles). The adaptive speed of FFD FFE is illustrated in Fig. 12(b), which shows the performance as a function of the training sequence length. The initial FFE coefficients were set to compensate 1080 km CD. It can be found that the FFE coefficients converged rapidly from the initial values to the optimal values during the first 10,000 bit (corresponding to 1 μs at 10 Gbit/s), and became steady thereafter. This suggests the potential of FFD FFE for applications in transparent optical networks where the reconfigurability of the add- and drop-nodes causes the transmission paths to vary rapidly.

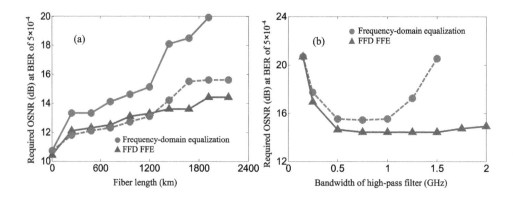

Fig. 11. (a) Required OSNR (dB) versus fiber length with optimized high-pass EF bandwidth. Solid and dashed lines represent the cases using 2 and 5 samples/bit. (b) Required OSNR versus the bandwidth of the high-pass EF. The fiber length is 2160 km. FFD FFE employs 2 samples/bit whilst the frequency-domain equalization uses 5 samples/bit.

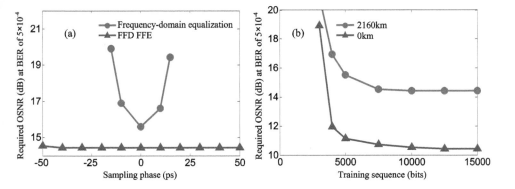

Fig. 12. Required OSNR (dB) versus (a) the sampling phase at 2160 km; (b) the training sequence length for FFD FFE at 2 samples/bit. The high-pass EF bandwidth is optimized.

4.3 FFD MLSE and comparison to DD MLSE, FFD FFE, and frequency-domain equalization

In Section 4.2, we verified that the well-known advantages of adaptive filters over fixed filters were clearly applicable to FFD EDC schemes, especially when the fixed filters were only designed to account for a restricted set of impairments. These advantages including improved tolerance to a wide high-pass filter bandwidth and the sampling phase misalignment also apply to MLSE. In this subsection, we will discuss the performance of FFD MLSE, and compare it with DD MLSE and other FFD-based EDC schemes, in particular, the adaptive FFD FFE. Fig. 13 shows the required OSNR as a function of fiber length using conventional DD MLSE (circles), FFD MLSE without a high-pass filter (triangles), and with a 1.25 GHz high-pass EF (squares). The full metric (Eq. (10)) was used and other system parameters were set to the optimal values as obtained in Section 4.1. The fiber nonlinearity and thermal noise were included. It was found that the FFD MLSE, without proper suppression of low-frequency amplification, performed worse than conventional DD MLSE (Bosco & Poggiolini, 2006; Savory et al., 2007) regardless of the memory length m. However, by optimizing the low-frequency response for the estimated frequency (squares), performance was significantly improved. This strongly suggests that systems based on FFD MLSE can offer greater reach than DD MLSE for both 4 and 16 states implementations. At a OSNR of 15 dB, the CD tolerance was enhanced from 270 km to 420 km, and from 400 km to 580 km for m of 2 and 4 respectively, representing approximately 50% performance improvement. More importantly, for optical networks with fixed transmission reach, FFD can greatly reduce the MLSE complexity when compared to DD, e.g. from 16 states to 4 states to achieve 400km. This improvement over DD MLSE has been experimentally verified recently (Zhao et al., 2010), where 4- and 16-state FFD MLSE was demonstrated to support 372 and 496 km BT Ireland's field-installed SMF respectively.

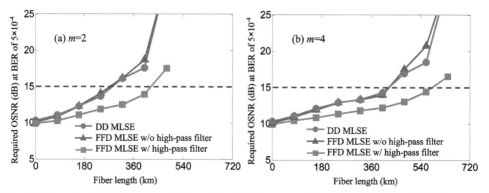

Fig. 13. Required OSNR versus distance for MLSE memory length m of (a) 2 and (b) 4.

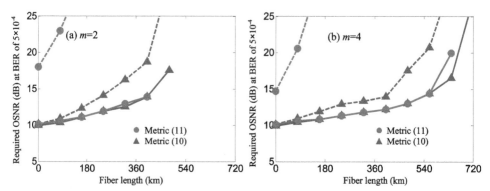

Fig. 14. Required OSNR as a function of transmission reach using metric (10) (triangles) and metric (11) (circles) for memory length m of (a) 2 and (b) 4. In (a) and (b), solid and dashed lines represent the cases with and without a 1.25 GHz high-pass filter.

As discussed in Section 3, the full metric (Eq. (10)) can be approximated by computing the two marginal probabilities instead of a single joint probability. Fig. 14 compares the required OSNR obtained using this reduced metric (11) (circles) and the full metric (10) (triangles) when the memory length m is (a) 2 and (b) 4. The figures clearly show that the high-pass filter was critical for the optimum operation of FFD MLSE using metric (11). This is because when the system was dominated by low-frequency amplification, a correlation in the noise statistics of the extracted real and imaginary components might be expected, so breaking the assumption leading to metric (11). In contrast, optimization of the low-frequency response enabled metric (11) to exhibit similar compensation performance to that using metric (10), with the advantage of a significant reduction in the complexity.

Having established that FFD MLSE based on the reduced metric (Eq. (11)) outperforms DD MLSE provided that an appropriate high-pass filter is employed in the phase estimation path, Fig. 15(a) compares the performance of two FFD-based adaptive compensation schemes, MLSE and FFE, using the optimized system parameters as discussed in Section 4.1.

It is clearly seen that 16-state (m=4) FFD MLSE exhibited better performance than FFD FFE with the same memory length, but had less compensation distance when compared to FFD FFE with increased memory lengths of m=8 and 16. Increasing the memory length of FFD MLSE can overcome this limitation but would increase the complexity exponentially, hindering its applications for longer-distance transmissions. However, for DCF free metro networks with distance around several hundred kilometers, FFD MLSE is a more effective approach. The reason is threefold. Firstly, it requires low implementation complexity for small m values ($\leq$4), which is achievable by modern microelectronic technologies. Secondly, it exhibits better performance limit than that of FFD FFE with the same m. Finally, FFD MLSE has much better tolerance to the noise and the associated noise amplification mechanisms in full-field reconstruction. Fig. 15(b) shows that when the system parameters were not fully optimized, the performance of FFD FFE was degraded severely, and was poorer than that of 16-state FFD MLSE even when m increased to 16. The curve using frequency-domain equalization (dotted line) was also depicted and exhibited the worst performance. Consequently, stringent limit on the design of system parameters should be placed on FFD FFE and frequency-domain equalization, but it can be greatly relaxed by FFD MLSE. This conclusion matches recent experimental demonstration (Zhao et al., 2010).

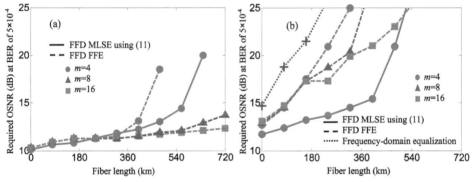

Fig. 15. Required OSNR versus fiber length by using metric (11) based FFD MLSE with memory length m of 4 (solid circles), and FFD FFE (dashed) with m of 4 (circles), 8 (triangles), and 16 (squares). (a): 8.5 GHz OBPF, 30 ps AMZI DTD, and 0 dBm optical power incident on the photodiodes; (b): 0.8 nm OBPF, 10 ps AMZI DTD, and -3 dBm optical power incident on the photodiodes. The dotted line in (b) represents frequency-domain equalization. In (a) and (b), the ER is 12 dB. Fiber nonlinear and thermal noise are included.

4.4 Discussion

Frequency-domain equalization is simple and cost-effective, but requires prior information of the dispersion experienced during the transmission. Although adaptation algorithms have been proposed for coefficient adaptation, this technique usually requires serial-to-parallel conversion into blocks for (inverse) Fourier transform (see Fig. 5), which would reduce the adaptation capability. On the other hand, FFD FFE improves the adaptation capability and can also equalize other linear impairments in addition to CD. Its complexity is approximately linearly proportional to the transmission distance and, for long-distance applications, is higher than the frequency-domain equalization method. Finally, the complexity of FFD MLSE increases exponentially with the transmission distance, hindering its applications for long-distance transmissions. However, for DCF free metro networks with transmission reach <500 km, FFD MLSE is a more effective approach. For long-distance applications (>500 km), the combination of the static frequency-domain equalization and adaptive FFD MLSE based on parametric channel estimation can well balance the complexity, performance, and adaptation speed, and will be investigated in the next section.

5. 10 Gbit/s OOK experiment for 0-900 km adaptive transmission

In this section, we experimentally demonstrate 10 Gbit/s OOK adaptive transmission for a wide range of distances from 0 to 900 km. The combination of static frequency-domain equalization and adaptive FFD MLSE with parametric channel estimation (Eq. (12)) was used in the experiment to balance the performance, complexity, and adaptation speed (Zhao & Ellis, 2011). Fig. 16 shows the experimental setup. A 1550 nm signal from a distributed feedback laser was intensity modulated using a MZM giving a 6 dB ER signal at 10 Gbit/s with 2^{15}-1 PRBS data. The OOK signal was transmitted over a re-circulating loop comprising 60 km of SMF with a signal launch power of -2.5 dBm per span. A 1 nm OBPF was used in the loop to suppress the ASE noise. At the receiver, the signal was detected with an optically pre-amplified receiver and a VOA was used to vary the input power to the EDFA. The preamplifier was followed by an OBPF with a 3 dB bandwidth of 0.3 nm, a second EDFA, and another OBPF with a 3 dB bandwidth of 0.8 nm. Then the optical signal was passed through an Kylia AMZI with 40 ps DTD and $\pi/2$ differential phase shift. The two outputs of the AMZI were detected by two 10 Gbit/s receivers. Both detected signals were simultaneously sampled by a real-time oscilloscope at 25 GSamples/s with 8-bit resolution. In off-line processing, an automatic algorithm was used to temporally align the signals from these two receiver chains, locate the position of the training sequence, and re-sample the signals. Note that due to the use of MLSE, the sampling phase was not strictly required to be at the eye centre. The received sequence was serial-to-parallel (S/P) converted to blocks with block size of 256 bits and 8-bit overlap between adjacent blocks for guard interval. Frequency-domain equalization was implemented based on block processing using (inverse) fast Fourier transform. The following FFD MLSE had 16 states and 2 samples/bit and used Gaussian based channel training (see Eq. (12)). 432,000 signal bits were processed.

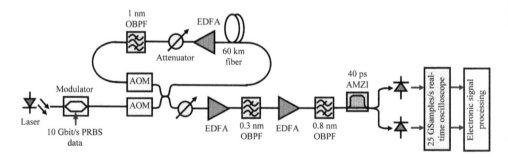

Fig. 16. Experimental setup. AOM: acoustic-optic modulator

Fig. 17(a) shows BER versus OSNR for 0, 480, 720, and 900 km. In this figure, the parameters of frequency-domain equalization were set to approximately fully compensate the CD, and the training time for the MLSE was 1 μs. Fig. 17(b) depicts the recovered eye diagrams after frequency-domain equalization for 900 km. The figure shows that the system operated well after 480 km and 720 km, with 3 dB and 4 dB OSNR penalty at BER of 10^{-3}, respectively. At 900 km, the slope was reduced due to non-ideally suppressed noise amplification. However, the best achievable BER was 1.5×10^{-4}, well below the forward error correction limit.

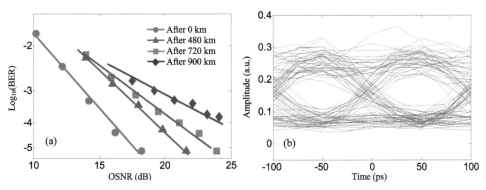

Fig. 17. (a) BER versus the OSNR. (b) eye diagram of the recovered signal after 900 km.

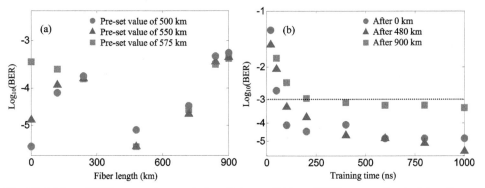

Fig. 18. (a) $\log_{10}$(BER) versus transmission distance. (b) $\log_{10}$(BER) versus the training time. The dotted line represents BER of 1×10^{-3}, used as the forward error correction limit.

Fig. 17 is based on the assumption that exact prior information of the fiber length has been obtained. In practice, this value may not be known and can also vary frequently over a wide range. Fig. 18(a) shows the performance when the frequency-domain equalization was preset to be a fixed value and MLSE was used to adaptively trim the impairments for various transmission distances. The training time of the MLSE was 1 μs and the received optical power into the pre-amplifier was -28 dBm. Note that the received OSNR was different for different transmission distances, with the case of 900 km exhibiting the worst OSNR of 23 dB. The figure shows that a BER better than 10^{-3} could be achieved for any measured distance up to 900 km when the pre-set value was between 500 km and 575 km. For the pre-set value beyond 575 km, the performance for short distances (<150 km) would be degraded due to the finite MLSE compensation window (~550 km at 16 states as shown Fig. 13(b)). This figure also implies that the system was insensitive to the exact pre-set dispersion value, so a coarse estimation was sufficient. To illustrate the adaptation speed of the system, Fig. 18(b) shows the BER versus the training time for three different distances when the frequency-domain equalization was pre-set to compensate 550 km CD. The figure shows that the performance converged rapidly during the first 200 ns for all distances. After 400 ns, the BER fell below 10^{-3} even for the longest distance, demonstrating the potential of FFD EDC in frequently configured optical networks.

6. Full-field detection for 40 Gbit/s offset DQPSK

In addition to amplitude-modulated OOK format, FFD can also be used in phase-modulated formats, which have been widely employed for 40 Gbit/s and beyond. In conventional differential quadrature phase shifted keying (DQPSK) system, at least two AMZIs and two pairs of balanced photodiodes are required for incoherent detection (Kikuchi et al., 2006; Liu & Wei, 2007). Furthermore, the near-zero intensity during a π phase shift between symbols limits the system performance unless complicated pre-distortion is used (Kikuchi & Sasaki, 2010). On the other hand, offset DQPSK format has been proposed in optical communications (Wree et al., 2004) to eliminate the near-zero intensity between symbols and this format exhibits the same spectral efficiency as conventional DQPSK. However, conventional offset DQPSK system has degraded receiver sensitivity and CD tolerance (Wree et al., 2004), which hinders its use for practical applications. In this section, we show that FFD based EDC can significantly improve the performance of the offset DQPSK system (Zhao & Ellis, 2011). The presented system uses a simpler pre-coder at the transmitter, only one AMZI, and one pair of photodiodes at the receiver, reducing the implementation cost when compared to conventional DQPSK. Consequently, it is promising for cost-sensitive 40 Gbit/s Ethernet or short metro networks.

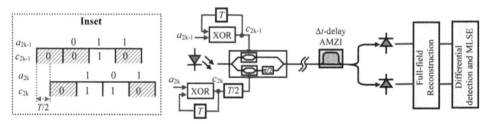

Fig. 19. Configuration of FFD-based offset DQPSK

Fig. 19 illustrates the configuration of FFD-based offset DQPSK. The transmitted data, a_k, is demultiplexed into the in-phase and quadrature tributaries, which are differentially encoded using exclusive OR (XOR) individually. Note that this pre-coder uses only two XOR gates and is much simpler than the conventional DQPSK pre-coder which typically requires the combination of >20 XOR, AND and NOT logic gates. The encoded quadrature signal is delayed by $T/2$ with respect to the in-phase signal (Inset in Fig. 19), where T is the symbol period. Consequently, the phase may possibly change every $T/2$, but each phase change can only be 0, $\pm\pi/2$. In addition, the possible zero intensity between symbols induced by instantaneous π phase shift in conventional DQPSK is eliminated. At the receiver, the optical front end and full-field reconstruction for offset-DQPSK are the same as those in the OOK format. However, an additional electrical-domain differential detection process is employed before the dispersion compensation stage, as depicted in Fig. 20. The performance of differential detection can be improved by exploiting the field differences between a symbol and its previous $(L-1)$ symbols, where $L>1$, resulting in better field reference. This method is conventionally implemented in the optical domain by using $(L-1)$ (or $2\times(L-1)$) AMZIs (Zhao & Chen, 2007). However, this implementation is complicated. By using FFD, multiple differential fields can be obtained in the electrical domain simply using delays and multiplications while only one optical AMZI is employed. In offset DQPSK, the phase may change every $T/2$, so two samples per symbol (or one sample per bit) are used. The multiple differential fields for the nth bit, $I_i(t_n)$, may be estimated by $conj(V_{full}(t_{n-i}))\cdot V_{full}(t_n)$, where i $(=1,...,L-1)$ denotes the ith branch of the differential field detection and $conj(\cdot)$ represents the conjugate. These differential samples are then fed into the MLSE. The metric of MLSE, $PM(a_n)$, used by the Viterbi algorithm to estimate the most likely transmitted data sequence, is given by:

$$PM(a_n) = PM(a_{n-1}) - \sum_{i=1}^{L-1} \log(p(\Re(I_i(t_n)), \Im(I_i(t_n)) \mid a_{n-m},...,a_n)) \tag{13}$$

where $\Re(I_i(t_n))$ (or $\Im(I_i(t_n))$) represents the real (or imaginary) part of the differential field $I_i(t_n)$. $p(\Re(I_i(t_n)), \Im(I_i(t_n))|a_{n-m},...,a_n)$ is the joint probability of the differential field given the transmitted data $a_{k-m},...,a_k$. m is the memory length. Eq. (13) shows that the size of the required lookup table for channel estimation and the complexity of metric computation scale approximately linearly with L. On the other hand, Viterbi decoding is independent of L and is the same as that in conventional MLSE.

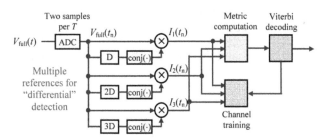

Fig. 20. Multiple-reference based differential detection and MLSE. D, 2D, and 3D represent one-, two-, and three-sample delay respectively.

Simulation implemented in Matlab was performed to verify the operating principle of this scheme. The analysis model was the same as Fig. 19. Two uncorrelated 20 Gbit/s data trains using $2^{11}-1$ pseudo-random binary sequence repeated nine times were differentially encoded individually. Each encoded data train generated an analogue electrical signal using raised-cosine shaped pulse with a roll-off coefficient of 0.4 and 40 samples per symbol. The response of the driving amplifier was 5th-order Bessel shaped with 20 GHz 3 dB bandwidth. The electrical signals were used to modulate a continuous wave light from a laser with 100 kHz linewidth. A piece of fiber with CD of 16 ps/km/nm was used to investigate the CD tolerance. At the receiver, the launch power into the preamplifier was adjusted to control the

OSNR. The preamplifier was followed by an OBPF with optimized bandwidth. The AMZI had a differential phase shift of $\pi/2$ and 10 ps DTD, unless otherwise stated. The signal power into the photodiodes was 3 dBm and the noise spectral power density of the photodiodes was 20 pA/Hz$^{1/2}$. After detection, the signals were amplified, filtered by a 30 GHz 4th-order Bessel EF, and processed as described above. MLSE had two samples per symbol, 5-bit resolution, and 16 states (considering two (or four) adjacent symbols (or bits)). The number of differential measurements used for metric computation, L, was varied from two to four. The simulation was iterated ten times with different random number seeds to give a total of 184,230 simulated symbols. The performance was evaluated using the required OSNR to achieve a BER of 1×10^{-3} by direct error counting.

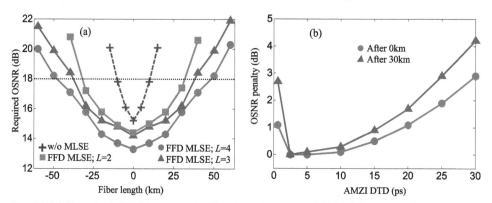

Fig. 21. (a) Required OSNR versus the fiber length without MLSE (pluses), and using 16-state MLSE with L of 2 (squares), 3 (triangles), and 4 (circles). The AMZI DTD is 10 ps. (b) OSNR penalty versus the AMZI DTD using 16-state MLSE and L=4. The OSNR penalty is defined as the penalty with respect to the OSNR value using optimized AMZI DTD.

Fig. 21(a) shows the performance of the offset DQPSK with and without MLSE. The OBPF bandwidth was optimized at the back-to-back case and the optimal value when using MLSE (16.5 GHz) was smaller than that without MLSE (23.5 GHz). In common with other MLSE investigation, this was due to the capability of MLSE to compensate filtering-induced ISI such that a narrow OBPF bandwidth could be used to mitigate the impact of the noise and the CD. The figure clearly depicts the benefit of MLSE with a larger number of differential measurements L. When using 16-state MLSE and L=4, a transmission distance of around ±50km could be supported for a required OSNR of 18dB (100km total dispersion tolerance range). Fig. 21(b) illustrates the low sensitivity of the system to the precise AMZI delay. Smaller DTDs gave more precise estimation of $V_f(t)$ and $V_p(t)$, and consequently resulted in reduced OSNR penalties. At 40 Gbit/s, less than 1 dB penalty was induced for an AMZI with DTD between 2.5 ps and 15 ps for both back-to-back and 30 km. Note that the DTD could not be reduced indefinitely due to the increased limit induced by thermal noise as discussed in Section 4.1.

7. Conclusions

FFD EDC, by surpassing the limited performance of current DD EDC products (≤300 km at 10 Gbit/s) and avoiding the high implementation cost of coherent detection EDC (for long-haul systems), is of particular value for applications in DCF-free transparent access/metro networks and Ethernet. For 10 Gbit/s metro networks with transmission reach of 300-500 km, FFD MLSE is an effective approach and can exhibit 50% performance improvement when compared to DD MLSE, or exponentially reduce the required state number for a fixed transmission reach. It is also more robust to non-optimized system parameters than full-field detection based frequency-domain equalization and FFE, and thus relaxes the system specifications. For transmission reaches longer than 500 km, the combination of cost-effective and static frequency-domain equalization and adaptive FFD MLSE with parametric channel estimation can obtain a balance of performance, complexity, and adaptation speed. 0-900 km adaptive transmission with less than 400ns adaptation time is achievable at 10 Gbit/s. For higher bit rate systems, FFD based offset DQPSK offers a cost-effective solution for 40 Gbit/s Ethernet or short metro networks, and when compared to conventional DQPSK with the same spectral efficiency, it uses a simpler pre-coder at the transmitter, only one AMZI and one pair of photodiodes at the receiver, while supporting ±50 km SMF transmission without optical compensation at 40 Gbit/s.

8. Acknowledgments

The authors acknowledge M.E. McCarthy from the Photonic Systems Group at the Tyndall National Institute for invaluable assistance with the experimental demonstration, and the contribution of D. Cassidy and W. McAuliffe from BT Ireland and P. Gunning from BT Innovate and Design for ongoing support. This work was financially supported by Science Foundation Ireland under grant number 06/IN/I969 and Enterprise Ireland under grant number CFTD/08/333.

9. References

Alfiad, M.; Van den Borne, D.; Napoli, A.; Koonen, A.M.J & De Waardt, H. (2008). A DPSK receiver with enhanced CD tolerance through optimized demodulation and MLSE. *IEEE Photonics Technology Letter*, vol. 20, no. 10, (May 2008), pp. 818-820.

Bosco, G. & Poggiolini, P. (2006); Long-distance effectiveness of MLSE IMDD receivers. *IEEE Photonics Technology Letters*, vol. 18, no. 9, (May 2006), pp. 1037-1039.

Bulow, H. & Thieleche, G. (2001). Electronic PMD mitigation – from linear equalization to maximum-likelihood detection. *Proceedings of Optical Fiber Communication (OFC) conference*, WAA3, Anaheim USA, March 2001.

Bulow, H.; Buchali, F. & Klekamp, A. (2008). Electronic dispersion compensation. *IEEE/OSA Journal of Lightwave Technology*, vol. 26, no. 1, (Jan. 2008), pp. 158-167.

Cai, Y. (2008). Coherent detection in long-haul transmission systems. *Proceedings of Optical Fiber Communication (OFC) conference*, OTuM1, San Diego USA, March 2008.

Chandrasekhar, S.; Gnauck, A.H.; Raybon, G.; Buhl, L.L.; Mahgerefteh, D.; Zheng, X.; Matsui, Y.; McCallion, K.; Fan, Z. & Tayebati, P. (2006). Chirp-managed laser and MLSE-RX enables transmission over 1200km at 1550nm in a DWDM environment in NZDSF at 10Gb/s without any optical dispersion compensation. *IEEE Photonics Technology Letters*, vol. 18, no. 14, (July 2006), pp. 1560-1562.

Ellermeyer, T.; Mullrich, J.; Rupeter, J.; Langenhagen, H.; Bielik, A. & Moller M. (2008). DA and AD converters for 25GS/s and above. *Proceedings of IEEE Summer Topical Meetings*, pp. 117-118, Acapulco Mexico, July 2008.

Ellis, A.D. & McCarthy, M.E. (2006). Receiver-side electronic dispersion compensation using passive optical field detection for low-cost 10Gbit/s 600km reach applications. *Proceedings of Optical Fiber Communication (OFC) conference*, OTuE4, Anaheim USA, March 2006.

Ellis, A.D., Zhao, J. & Cotter, D. (2010). Approaching the Non-linear Shannon Limit. *IEEE Journal of Lightwave Technology*, vol. 28, no. 4, (Feb. 2010), pp 423-433.

Farbert, A.; Langenbach, S.; Stojanovic, N.; Dorschky, C.; Kupfer, T.; Schulien, C.; Elbers, J.-P.; Wernz, H.; Griesser, H. & Glingener, C. (2004). Performance of a 10.7Gb/s receiver with digital equalizer using maximum likelihood sequence estimation. *Proceedings of European Conference on Optical Communications*, PDP Th4.1.5, Stockholm Sweden, Sep. 2004.

Franceschini, M.; Bongiorni, G.; Ferrari, G.; Raheli, R.; Meli, F. & Castoldi, A. (2007). Fundamental limits of electronic signal processing in direct-detection optical communications. *IEEE/OSA Journal of Lightwave Technology*, vol. 25, no. 7, (July 2007), pp. 1742-1753.

Gene, J.M.; Winzer, P.J.; Essiambres, R.J.; Chandrasekhar, S.; Painchaud, Y. & Guy, M. (2007). Experimental study of MLSE receivers in the presence of narrowband and vestigial sideband optical filtering. *IEEE Photonics Technology Letters*, vol. 19, no. 16, (Aug. 2007), pp. 1224-1227.

Haunstein, H. & Urbansky, R. (2004). Application of electronic equalization and error correction in lightwave systems. *Proceedings of European Conference on Optical Communications*, Th1.5.1, Stockholm Sweden, Sep. 2004.

Iwashita, K. & Takachio, N. (1988). Compensation of 202km single-mode fiber chromatic dispersion in 4Gbit/s optical CPFSK transmission experiment. Electronics Letters, vol. 24, no. 12, (June 1988), pp. 759-760.

Jeruchim, M.C. (1984). Techniques for estimating the bit error rate in the simulation of digital communication systems. *IEEE Journal of Selected Areas in Communications*, vol. SAC-2, no. 1, (Jan. 1984), pp. 153-170.

Kikuchi, N.; Mandai, K.; Sasaki, S. & Sekine, K. (2006). Proposal and first experimental demonstration of digital incoherent optical field detector for chromatic dispersion compensation. *Proceedings of European Conference on Optical Communications*, PDP Th4.4.4, Cannes France, Sep. 2006.

Kikuchi, N. & Sasaki, S. (2010). Improvement of chromatic dispersion and differential group delay tolerance of incoherent multilevel signalling with receiver-side digital signal processing. *Proceedings of Optical Fiber Communication (OFC) conference*, OWV6, San Diego USA, March 2010.

Liu, X. & Wei, X. (2007). Electronic dispersion compensation based on optical field reconstruction with orthogonal differential direct-detection and digital signal processing. *Proceedings of Optical Fiber Communication (OFC) conference*, OTuA6, Anaheim USA, March 2007.

Liu, X.; Chandrasekhar, S. & Leven, A. (2008). Digital self-coherent detection. *Optics Express*, vol. 16, no. 2, (Jan. 2008), pp. 792-803.

McCarthy, M.E. & Ellis, A.D. (2007). Electronic Dispersion Compensation Utilising Full Optical Field Estimation. *Mediterranean Journal of Electronics and Communications*, vol 3, no. 4, (Oct. 2007), pp. 144-151.

McCarthy, M.E.; Zhao, J.; Ellis, A.D. & Gunning, P. (2008). Full field receiver side processing for electronic dispersion compensation. *Proceedings of IEEE Summer Topical Meetings*, pp. 171-172, Acapulco Mexico, July 2008.

McCarthy, M.E.; Zhao, J.; Gunning, P. & Ellis, A.D. (2008). A novel field-detection maximum-likelihood sequence estimation for chromatic dispersion compensation. *Proceedings of European Conference on Optical Communications*, We.2.E.5, Brussels Belgium, Sep. 2008.

McCarthy, M.E.; Zhao, J.; Ellis, A.D. & Gunning, P. (2009). Full-field electronic dispersion compensation of 10Gbit/s OOK signal over 4×124km field-installed signal-mode fiber. *IEEE/OSA Journal of Lightwave Technology*, vol. 27, no. 23, (Dec. 2009), pp. 5327-5334.

McGhan, D.; Laperle, C.; Savchenko, A.; Li, C; Mak, G. & O'Sullivan, M. (2005). 5120km RZ-DPSK transmission over G652 fiber at 10Gbit/s with no optical dispersion compensation. *Proceedings of Optical Fiber Communication (OFC) conference*, PDP27, Anaheim USA, March 2005.

McGhan, D.; O'Sullivan, M.; Sotoodeh, M.; Savchenko, A.; Bontu, C.; Belanger, M. & Roberts K. (2006). Electronic dispersion compensation. *Proceedings of Optical Fiber Comunication (OFC) conference*, OWK1, Anaheim USA, March 2006.

McNicol, J.; O'Sullivan, M.; Roberts, K.; Comeau, A.; McGhan, D. & Strawczynski, L. (2005). Electronic domain compensation of optical dispersion. *Proceedings of Optical Fiber Communication (OFC) conference*, OThJ3, Anaheim USA, March 2005.

Poggiolini, P.; Bosco, G.; Benlachtar, Y.l; Savory, S.J.; Bayvel, P.; Killey, R.I. & Prat, J. (2008). Long-haul 10Gbit/s linear and nonlinear IMDD transmission over uncompensated standard fiber using a SQRT-metric MLSE receiver. *Optics Express*, vol. 16, no. 17, (Aug. 2008), pp. 12919-12936.

Polley, A. & Ralph, S.E. (2007). Receiver-side adaptive opto-electronic chromatic dispersion compensation. *Proceedings of Optical Fiber Communication (OFC) conference*, JThA51, Anaheim USA, March 2007.

Proakis, J.G. (2000). *Digital communications*. McGraw-Hill, New York.

Savory, S.J.; Benlachtar, Y.; Killey, R.I.; Bayvel, P.; Bosco, G.; Poggiolini, P.; Prat, J. & Omella, M. (2007). IMDD transmission over 1040km of standard single-mode fiber at 10Gbit/s using a one-sample-per-bit reduced complexity MLSE receiver. *Proceedings of Optical Fiber Communication (OFC) conference*, OThK2, Anaheim USA, March 2007.

Savory, S.J.; Gavioli, G.; Killey, R.I. & Bayvel, P. (2007). Electronic compensation of chromatic dispersion using a digital coherent receiver. *Optics Express*, vol. 15, no. 5, (March 2007), pp. 2120-2126.

Schube, S. & Mazzini, M. (2007). Testing and interoperability of 10GBASE-LRM optical interfaces. *IEEE Communication Magazine*, vol. 45, (March 2007), pp. s26-s31.

Taylor, M.G. (2004). Coherent detection method using DSP for demodulation of signal and subsequent equalization of propagation impairments. *IEEE Photonics Technology Letters*, vol. 16, no. 2, (Feb. 2004), pp. 674-676.

Tsukamoto, S.; Katoh, K. & Kikuchi, K. (2006). Unrepeated transmission of 20Gb/s optical quadrature phase-shift-keying signal over 200km SMF based on digital processing of homodyne-detected signal for group-velocity dispersion compensation. *IEEE Photonics Technology Letters*, vol. 18, no. 9, (May 2006), pp. 1016-1018.

Weem, J.; Kirkpatrick, P. & Verdiell J. (2005). Electronic dispersion compensation for 10Gigabit communication links over FDDI legacy multimode fiber. *Proceedings of Optical Fiber Communication (OFC) conference*, OFO4, Anaheim USA, March 2005.

Winters, J.; Gitlin, R. (1990). Electrical signal processing techniques in long-haul fiber-optic systems. IEEE Transactions on Communications, vol. 38, no. 9, (Sep. 1990), pp. 1439-1453.

Wree, C.; Serbay, M.; Leibrich, J. & Rosenkranz, W. (2004). Offset-DQPSK modulation format for 40Gb/s and comparison to RZ-DQPSK in WDM environment. *Proceedings of Optical Fiber Communication (OFC) conference*, MF62, Anaheim USA, March 2004.

Xia, C. & Rosenkranz, W. (2006). Electronic dispersion compensation for different modulation formats with optical filtering. *Proceedings of Optical Fiber Communication (OFC) conference*, OWR2, Anaheim USA, March 2006.

Zhao, J. & Chen, L.K. (2007). Three-chip DPSK maximum likelihood sequence estimation for chromatic dispersion and polarization-mode dispersion compensation. *Optics Letters*, vol. 32, no. 12, (June 2007), pp. 1746-1748.

Zhao. J.; McCarthy, M.E. & Ellis, A.D. (2008). Electronic dispersion compensation using full optical field reconstruction in 10Gbit/s OOK based systems. *Optics Express*, vol. 16, no. 20, (Sep. 2008), pp. 15353-15365.

Zhao, J.; McCarthy, M.E.; Gunning, P. & Ellis, A.D. (2009). Mitigation of pattern sensitivity in full-field electronic dispersion compensation. *IEEE Photonics Technology Letters*, vol. 21, no. 1, (Jan. 2009), pp. 48-51.

Zhao, J.; McCarthy, M.E.; Gunning, P. & Ellis, A.D. (2010). Simplified field reconstruction and adaptive system optimization in full-field FFE. *Proceedings of Optical Fiber Communication (OFC) conference*, OWV7, San Diego USA, March 2010.

Zhao, J.; McCarthy, M.E.; Gunning, P. & Ellis, A.D. (2010). Chromatic dispersion compensation using full-field maximum likelihood sequence estimation. *IEEE/OSA Journal of Lightwave Technology*, vol. 28, no. 7, (April 2010), pp. 1023-1031.

Zhao, J.; Bessler, V. & Ellis, A.D. (2010). Full-field feed-forward equalizer with adaptive system optimization. *Optical Fiber Technology*, vol. 16, no. 5, (Oct. 2010), pp. 323-328.

Zhao, J. & Ellis, A.D. (2011). Demonstration of 10Gbit/s transmission over 900km SMF with <400ns adaptation time using full-field EDC. *IEE Electronics Letters*, vol. 47, no. 12, (June 2011), pp. 711-712.

Zhao, J. & Ellis, A.D. (2011). Full-field detection based multi-chip MLSE for offset-DQPSK modulation format. *Proceedings of European Conference on Optical Communications*, Tu.5.A.2, Geneva, Switzerland, Sep. 2011.

3

Effect of Clear Atmospheric Turbulence on Quality of Free Space Optical Communications in Western Asia

Abdulsalam Alkholidi and Khalil Altowij

Faculty of Engineering, Electrical Engineering Department, Sana'a University, Sana'a, Yemen

1. Introduction

The Free Space Optical (FSO) communication is also known as Wireless Optical Communication (WOC), Fibreless, or Laser Communication (Lasercom). FSO communication is one of the various types of wireless communication which witnesses a vast development nowadays. FSO provides a wide service and requires point-to-point connection between transmitter and receiver at clear atmospheric conditions. FSO is basically the same as fiber optic transmission. The difference is that the laser beam is collimated and sent through atmosphere from the transmitter, rather than guided through optical fiber [1]. The FSO technique uses modulated laser beam to transfer carrying data from a transmitter to a receiver. FSO is affected by attenuation of the atmosphere due to the instable weather conditions. Since the atmosphere channel, through which light propagates is not ideal.

In this study we will take republic of Yemen as case study. In some mountainous areas in Yemen, it is difficult to install the technique of fiber optics. But FSO technique will solve this problem with same proficiency and quality provided by fiber optics. FSO systems are sensitive to bad weather conditions such as *fog, haze, dust, rain and turbulence*. All of these conditions act to attenuate light and could block the light path in the atmosphere. As a result of these challenges, we have to study weather conditions in detail before installing FSO systems. This is to reduce effects of the atmosphere also to ensure that the transmitted power is sufficient and minimal losses during bad weather.

This chapter aims to study and analyze the atmosphere effects on the FSO system propagation in the Republic of Yemen weather environment. The study is focused more on the effects of haze, rain and turbulence on the FSO systems.

The analysis conducted depends basically on statistical data of the weather conditions in Yemen obtained from the Civil Aviation and Meteorology Authority (CAMA) for visibility and wind velocity and from the Public Authority for Water Resources (PAWR) for the rainfall rate intensity. So, the prominent objectives of this work are:

1. Calculating the scattering coefficient, atmospheric attenuation, total attenuation in the hazy and rainy days and scintillation at the clear days.

2.   Studying the performance of FSO system at wavelengths 780 nm, 850 nm and 1550 nm, beam divergence angle, transmitter and receiver diameter apertures and transmission range.

The scope of this chapter focuses on studying and analyzing of FSO propagation under weather conditions in Yemen environment for outdoor system. The atmospheric effects divided into two kinds: atmospheric attenuation and atmospheric turbulence. Atmospheric attenuation due to Mie scattering is related to the haze and it is a function of the visibility, and attenuation due to rainfall independent on wavelength. The atmospheric turbulence is due to variance of refractive index structure.

There are three factors which enable us to test the FSO performance as: design, uncontrollable and performance. Design factors are relating to FSO design such as light power, wavelength, receiver and transmitter aperture diameter, link range and detector sensitivity. Uncontrollable elements such as rainfall elements include rainfall rate and raindrop radius, haze element include visibility and turbulence element include refractive index structure. Performance of system was tested during the rainy days and hazy days which can be calculated from the effect of scattering coefficient, atmospheric attenuation and total attenuation. However, the system performance in the clear days can be calculated from the effect of variance.

The remainder of this chapter is organized as follow: Discuses briefly the background of FSO. The concept of and the several stages of FSO transceiver are explained briefly. Illustrate the losses of FSO system due to atmosphere channel and geometric losses. Defining the aerosols and visibility and how they effect on FSO system were introduced. Beer's law which describes attenuation of atmospheric channel due to absorption and scattering coefficient was introduced. The atmospheric attenuation types are explained. Stroke's law which describes scattering coefficient due to rainfall was introduced. Geometrical loss and total attenuation are discussed. Represent the analytical study of Yemeni environment. Scattering coefficient and attenuation in hazy days at average and low visibility at three wavelengths (780 nm, 850 nm and 1550 nm) were introduced. The atmospheric attenuation at the lowest visibility conditions was plotted with the link range. Scattering coefficient and atmospheric attenuation in rainy days were plotted once versus rainfall rate and once again versus raindrop radius. Scattering coefficient and atmospheric attenuation to haze in Sana'a, Aden and Taiz cities were calculated. The geometric loss for two commonly used designs of transceivers was evaluated. The conclusion is done based on the overall findings of this work.

2. Fundamental of free space optical communication

FSO is a technique used to convey data carried by a laser beam through the atmosphere. While FSO offers a broadband service, it requires Lone of Sight (LOS) communication between the transmitter and receiver as shown in the Fig. (1)[1].

The atmosphere has effects on the laser beam passing through it, so the quality of data received is affected. To reduce this effect, the fundamental system components must be designed to adopt with the weather conditions. This design is mostly related to transmitter and receiver components. In the following subsection, we will tackle discuss the components and the basic system of FSO.

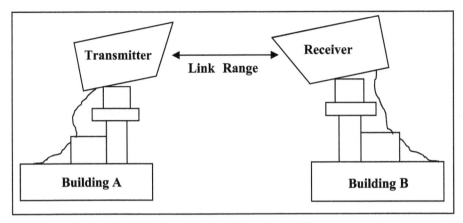

Fig. 1. Schematic showing FSO Transmitter and Receiver LOS.

2.1 FSO communication subsystem

FSO communication is a line of sight technology that uses laser beam for sending the very high bandwidth digital data from one point to another through atmosphere. This can be achieved by using a modulated narrow laser beam lunched from a transmission station to transmit it through atmosphere and subsequently received at the receiver station. The generalized FSO system is illustrated in Fig. (2), it is typically consists of transmitter, FSO channel and a receiver.

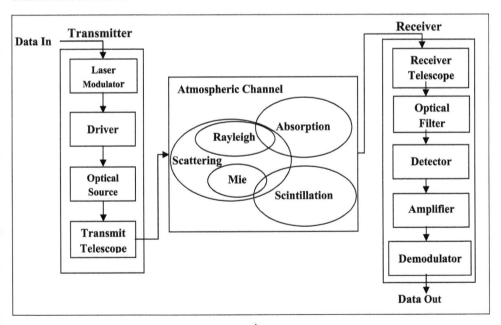

Fig. 2. Block Diagram of a Terrestrial FSO System.

a. Transmitter

Transmitter transforms the electrical signal to an optical signal and it modulates the laser beam to transfer carrying data to the receiver through the atmosphere channel. The transmitter consists of four parts as shown in Fig. (2): laser modulator, driver, optical source and transmit telescope.

- **Laser modulator**

Laser modulation means the data were carried by a laser beam. The modulation technique can be implemented in following two common methods: internal modulation and external modulation [2].

Internal modulation: is a process which occurs inside the laser resonator and it depends on the change caused by the additive components and change the intensity of the laser beam according to the information signal.

External modulation: is the process which occurs outside the laser resonator and it depends on both the polarization phenomena and the refractive dualism phenomenon.

- **Driver**

Driver circuit of a transmitter transforms an electrical signal to an optical signal by varying the current flow through the light source.

- **Optical source**

Optical source may be a laser diode (LD) or light emitting diode (LED), which used to convert the electrical signal to optical signal.

A laser diode is a device that produces optical radiation by the process of stimulated emission photons from atoms or molecules of a lasing medium, which have been excited from a ground state to a higher energy level. A laser diode emits light that is highly monochromatic and very directional. This means that the LD's output has a narrow spectral width and small output beam angle divergence. LDs produce light waves with a fixed-phase relationship between points on the electromagnetic wave. There are two common types of laser diode: Nd:YAG solid state laser and fabry-perot and distributed-feedback laser (FP and DFB) [3].

- **Laser source selection criteria for FSO**

The selection of a laser source for FSO applications depends on various factors. They factors can be used to select an appropriate source for a particular application. To understand the descriptions of the source performance for a specific application, one should understand these detector factors. Typically the factors that impact the use of a specific light source include the following [4]:

- Price and availability of commercial components
- Transmission power and lifetime
- Modulation capabilities
- Eye safety
- Physical dimensions and compatibility with other transmission media.

- **Transmitter telescope**

The transmitter telescope collects, collimates and directs the optical radiation towards the receiver telescope at the other end of the channel.

b.   FSO channel

For FSO links, the propagation medium is the atmosphere. The atmosphere may be regarded as series of concentric gas layers around the earth. Three principal atmospheric layers are defined in the homosphere [5], the troposphere, stratosphere and mesosphere. These layers are differentiated by their temperature gradient with respect to the altitude. In FSO communication, we are especially interested in the troposphere because this is where most weather phenomena occur and FSO links operate at the lower part of this layer [5].

The atmosphere is primarily composed of nitrogen (N_2, 78%), oxygen (O_2, 21%), and argon (Ar, 1%), but there are also a number of other elements, such as water (H_2O, 0 to 7%) and carbon dioxide (CO_2, 0.01 to 0.1%), present in smaller amounts. There are also small particles that contribute to the composition of the atmosphere; these include particles (aerosols) such as haze, fog, dust, and soil [6].

Propagation characteristics of FSO through atmosphere drastically change due to communication environment, especially, the effect of weather condition is strong. The received signal power fluctuates and attenuates by the atmospheric obstacles such as rain, fog, haze and turbulence in the propagation channel. The atmospheric attenuation results from the interaction of the laser beam with air molecules and aerosols along the propagation. The main effects on optical wireless communication are absorption, scattering, and scintillation [7].

c.   Receiver

The receiver optics consists of five parts as shown in Fig. 2: receiver telescope, optical filter, detector, amplifier and demodulator.

- **Receiver telescope**

The receiver telescope collects and focuses the incoming optical radiation on to the photo detector. It should be noted that a large receiver telescope aperture is desirable because it collects multiple uncorrelated radiation and focuses their average on the photo detector [8].

- **Optical filter**

By introducing optical filters that allow mainly energy at the wavelength of interest to impinge on the detector and reject energy at unwanted wavelengths, the effect of solar illumination can be significantly minimized [6].

- **Detector**

The detector also called photodiode (PD) is a semiconductor devices which converts the photon energy of light into an electrical signal by releasing and accelerating current conducting carriers within the semiconductors. Photodiodes operate based on photoconductivity principals, which is an enhancement of the conductivity of p-n semiconductor junctions due to the absorption of electromagnetic radiation. The diodes are generally reverse-biased and capacitive charged [9]. The two most commonly used photodiodes are the pin photodiode and the avalanche photodiode (APD) because they have good quantum efficiency and are made of semiconductors that are widely available commercially [10].

- **Features of detector**

The performance characteristics indicate how a detector responds to an input of light energy. They can be used to select an appropriate detector for a particular application. To understand the descriptions of detector performance and to be able to pick a detector for a specific application, one should understand these detector characteristics. In general, the following properties are needed:

- A high response at the wavelength to be detected.
- A small value for the additional noise is introduced by the detector.
- Sufficient speed of response.

2.2 FSO system

FSO system refers to the transmission of modulated visible or infrared (IR) beams through the atmosphere to obtain broadband communications. This technique requires clear line of site between the transmitter and the receiver. FSO system provides higher bandwidth at faster speed. The elements of FSO designed which must be considered by a prudent user are wavelength, beam divergence angle, aperture diameter and range.

2.2.1 Wavelength

To select the best wavelength to use for free-space optical communication systems, you must consider several factors, such as availability of components, eye safety considerations, required transmission distance, price, and so on. The availability of components is light sources and detectors [4]. Eye safety is one of the most important restrictions to the optical power level emitted by a wireless IR transmitter. Lasers of much higher power can be used more safely with 1550 nm systems than with 850 nm and 780 nm systems. This is because wavelengths is less than about 1400 nm focused by the human cornea into a concentrated spot falling on the retina as shown in Fig. (3), which can cause eye damage.

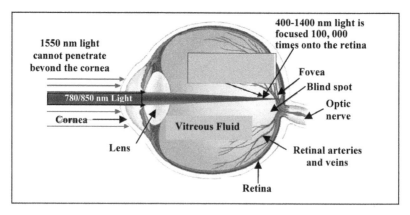

Fig. 3. Penetration of Light into Eyeball.

The allowable safe laser power is about 50 times higher at 1550 nm. This factor, 50 is important as it provides up to 17 dB additional margin, allowing the system to propagate over longer distances, through heavier attenuation, and to support higher data rates [11]. However, 1550 nm systems are at least 10 times more expensive than 850 nm systems [16]. The highest data rate available with commercial 850 nm systems is 622 Mbps, and 2.5 Gbps for 1550 nm systems. Table (1) illustrates the maximum Permissible Exposure (MPE) limited for unaided viewing, in case of 850 nm and 1550 nm [11].

Wavelength	Maximum Permissible Exposure (MPE)
850 nm	2 mW/cm²
1550 nm	100 mW/cm²

Table 1. Maximum Permissible Exposure Limited for "unaided viewing".

2.2.2 Beam divergence

Beam divergence purposely allows the beam to diverge or spread. The advantage using narrow beam in FSO system generates much higher data rates and increases the security. Laser generated with extreme narrow light can be easily modulated with voice and data information. The beam spread is dependent on the beam divergence angle and transmission range. Typically, 1 mrad to 8 mrad beam divergence spreads 1 to 8 m at distance of 1 km. To avoid spreading of a large beam, it is better to use narrow beam divergence such as 1 mrad [12-14].

2.2.3 Aperture diameter

In FSO system a smaller diameter of transmitter and a larger diameter of receiver aperture are needed to establish high data rate communication links. The diameter aperture of the transmitter and receiver must be adequate for the weather conditions. When the laser beam propagates through atmosphere, the beam is spreading, at a distance L from the source, due to the turbulence. If the turbulence cell is larger than the beam diameter, and the diameter of receiver aperture is small, then the beam bends and it can cause the signal to complete missing the received unit. A large size of diameter aperture of receiver is able to reduce turbulence effect on FSO [12][15]. Two particular design specifications are made in Table (2) due to particular implementation especially based on the existing product available in the industry [12].

Design	Diameter of transmitter aperture	Diameter of receiver aperture
Design 1	18 cm	18 cm
Design 2	3.5 cm	20 cm

Table 2. Diameter of Transmitter and Receiver Aperture of an FSO System.

2.2.4 Range

Distance between a transmitter and a receiver impacts the performance of FSO systems in three ways. First, even in clear weather conditions such as scintillation, the beam diverges and the detector element receives less power. Second, the total transmission loss of the beam increases with increasing distance. Third, scattering and absorption effect accumulates with longer distances. Therefore, the value for the scintillation fading margin in the overall power budget will increase to maintain a predefined value for the BER. Most commercially available FSO systems are rated for operation between 25–5000 m, with high-powered military and satellite systems capable of up to 2000 km. Most systems rated for greater than 1 km incorporate three or more lasers operating in parallel to mitigate distance related to issues. It is interesting to note that in the vacuum of space, FSO can achieve distances of thousands of kilometers [4].

3. Formulations

Free space optical communication link requires a good understanding of the atmosphere as the laser beam has to propagate through it. The atmosphere not only attenuates the light wave but also distorts and bends it. Attenuation is primarily the result of absorption and scattering by molecules and particles (aerosols) suspended in the atmosphere. Distortion, on the other hand, is caused by atmospheric turbulence due to index of refraction fluctuations. Attenuation affects the mean value of the received signal in an optical link whereas distortion results in variation of the signal around the mean. Often, atmospheric attenuation can be the limiting factor in an optical communication link through the atmosphere [16].

3.1 Aerosols

Aerosols are particles suspended in the atmosphere such as fog, haze, dust and smog, and they have diverse nature, shape, and size. Aerosols can vary in distribution, constituents, and concentration. The larger concentration of aerosols is in the boundary layer (a layer up to 2 km above the earth surface). Above the boundary layer, aerosol concentration rapidly decreases [3].

Scattering is the main interaction between aerosols and a propagating beam. Because the sizes of the aerosol particles are comparable to the wavelength of interest in optical communications, Mie scattering theory is used to describe aerosol scattering [17].

Type	Radius (μm)	Concentration (in cm^{-3})
Air molecules	10^{-4}	10^{19}
Aerosol	10^{-2} to 1	10 to 10^3
Fog	1 to 10	10 to 100
Cloud	1 to 10	100 to 300
Raindrops	10^2 to 10^4	10^{-5} to 10^{-2}
Snow	10^3 to 5×10^3	N/A
Hail	5×10^3 to 5×10^4	N/A

Table 3. Radius Ranges for Various Types of Particles.

Such a theory specifies that the scattering coefficient of aerosols is a function of the aerosols, their size distribution, cross section, density, and wavelength of operation. The different types of atmospheric constituents' sizes and concentrations of the different types of atmospheric constituents are listed in Table (3) [6], [18].

3.2 Visibility runway visual range (RVR)

Visibility defined as (Kruse model) means of the length where an optical signal of 550 nm is reduced to 0.02 of its original value. It is characterized by the transparency of the atmosphere, estimated by a human observer. Visibility is a useful measure of the atmosphere containing fog, smog, dust, haze, mist, clouds and other contaminating particles. Thick fog can reduce visibility down to a few meters; and maritime mist and clouds can affect visibility in the same way [19].

Low visibility will decrease the effectiveness and availability of FSO systems, and it can occur during a specific time period within a year or at specific times of the day. Low visibility means the concentration and size of the particles are higher compared to average visibility. Thus, scattering and attenuation may be caused more in low visibility conditions [20]. Attenuation can reach hundreds of dB per km for low visibility values, and is higher at shorter wavelength [21]. Low visibility and the associated high scattering coefficients are the most limiting factors for deploying FSO systems over longer distances [4].

3.3 Atmospheric attenuation

Atmospheric attenuation is defined as the process whereby some or all of the electromagnetic wave energy is lost when traversing the atmosphere. Thus, atmosphere causes signal degradation and attenuation in a FSO system link in several ways, including absorption, scattering, and scintillation. All these effects are time-varying and will depend on the current local conditions and weather. In general, the atmospheric attenuation is given by the following Beer's law Eq. (1) [22]:

$$\tau = \exp(-\beta L) \tag{1}$$

Where:

τ: is the atmospheric attenuation.

β: is the total attenuation coefficient given as:

$$\beta = \beta_{abs} + \beta_{scat} \tag{2}$$

L: is the distance between (T_x) and (R_x) in kilometer.

β_{abs}: is the molecular and aerosol absorption.

β_{scat}: is the molecular and aerosol scattering.

3.4 Absorption

Absorption is caused by the beam's photons colliding with various finely dispersed liquid and solid particles in the air such as water vapor, dust, ice, and organic molecules. The aerosols that have the most absorption potential at infrared wavelengths include water, O_2, O_3, and CO_2. Absorption has the effect of reducing link margin, distance and the availability of the link [4].

The absorption coefficient depends on the type of gas molecules, and on their concentration. Molecular absorption is a selective phenomenon which results in the spectral transmission of the atmosphere presenting transparent zones, called atmospheric transmission windows [5], and shown in Fig. (4), which allows specific frequencies of light to pass through it. These windows occur at various wavelengths. The atmospheric windows due to absorption are created by atmospheric gases, but neither nitrogen nor oxygen, which are two of the most abundant gases, contribute to absorption in the infrared part of the spectrum [6].

It is possible to calculate absorption coefficients from the concentration of the particle and the effective cross section such as shown in Eq. (3) [2]:

$$\beta_{abs} = \alpha_{abs} N_{abs} \left[\frac{1}{km}\right] \tag{3}$$

Where:

α_{abs}: is the effective cross section of the absorption particles [km^2].

N_{abs}: is the concentration of the absorption particles [1/km^3].

The absorption lines at visible and near infrared wavelengths are narrow and generally well separated. Thus, absorption can generally be neglected at wavelength of interest for free space laser communication [22]. Another reason for ignoring absorption effect is to select wavelengths that fall inside the transmittance windows in the absorption spectrum [23].

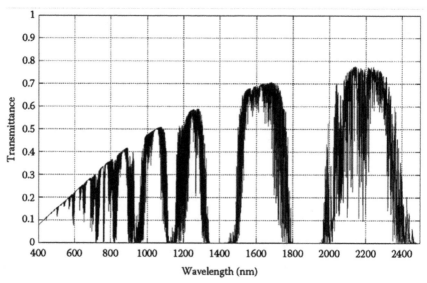

Fig. 4. Atmospheric Transmittance Window with Absortion Contribution.

3.5 Scattering

Scattering is defined as the dispersal of a beam of radiation into a range of directions as a result of physical interactions. When a particle intercepts an electromagnetic wave, part of the wave's energy is removed by the particle and re-radiated into a solid angle centered at it. The scattered light is polarized, and of the same wavelength as the incident wavelength, which means that there is no loss of energy to the particle [19].

There are three main types of scattering: (1) Rayleigh scattering, (2) Mie scattering, and (3) non-selective scattering. The scattering effect depends on the characteristic size parameter x_0, such as that $x_0 = 2\pi r / \lambda$, where, r is the size of the aerosol particle encountered during propagation [24]. If $x_0 << 1$, the backward lobe becomes larger and the side lobes disappear[25] and the scattering process is termed as Rayleigh scattering. If $x_0 \approx 1$, the backward lobe is symmetrical with the forward lobe and then it is Mie scattering. For $x_0 >> 1$, the particle presents a large forward lobe and small side lobes that start to appear and the scattering process is termed as non-selective scattering. It is possible to calculate the scattering coefficients from the concentration of the particles and the effective cross section such as Eq. (4) [2]:

$$\beta_{scat} = \alpha_{scat} N_{scat} [1/km] \tag{4}$$

Where:

β_{scat}: is either Rayleigh (molecular) β_m or Mie (aerosols) β_a scattering.

α_{scat}: is a cross-section parameter $[km^2]$.

N_{scat}: is a particle concentration $[1/km^3]$.

The total scattering can be written as presented in Eq. (5):

$$\beta_{scat} = \beta_m + \beta_a [1/km] \tag{5}$$

3.5.1 Rayleigh (molecular) scattering

Rayleigh scattering refers to scattering by molecular and atmospheric gases of sizes much less than the incident light wavelength. The Rayleigh scattering coefficient is given by Eq. (6) [2]:

$$\beta_m = \alpha_m N_m [1/km] \tag{6}$$

Where:

α_m: is the Rayleigh scattering cross-section $[km^2]$.

N_m: is the number density of air molecules $[1/km^3]$.

Rayleigh scattering cross section is inversely proportional to fourth power of the wavelength of incident beam (λ^{-4}) as the following relationship:

$$\alpha_m = \frac{8\pi^3 (n^2-1)^2}{3N^2 \lambda^4} [km^2] \tag{7}$$

Where:

n: is the index of refraction.

λ: is the incident light wavelength $[m]$.

N: is the volumetric density of the molecules $[1/km^3]$.

The result is that Rayleigh scattering is negligible in the infrared waveband because Rayleigh scattering is primarily significant in the ultraviolet to visible wave range [19].

3.5.2 Mie (aerosols) scattering

Mie scattering occurs when the particle diameter is equal or larger than one-tenth the incident laser beam wavelength, Mie scattering is the main cause of attenuation at laser wavelength of interest for FSO communication at terrestrial altitude. Transmitted optical beams in free space are attenuated most by the fog and haze droplets mainly due to dominance of Mie scattering effect in the wavelength band of interest in FSO (0.5 μm – 2 μm). This makes fog and haze a keys contributor to optical power/irradiance attenuation. The attenuation levels are too high and obviously are not desirable [26].

The attenuation due to Mie scattering can reach values of hundreds of dB/km [27], [24] (with the highest contribution arising from fog). The Mie scattering coefficient expressed as follows, see Eq. (8) [2]:

$$\beta_a = \alpha_a N_a [1/km] \tag{8}$$

Where:

α_a: is the Mie scattering cross-section $[km^2]$.

N_a: is the number density of air particles $[1/km^3]$.

An aerosol's concentration, composition and dimension distribution vary temporally and spatially varying, so it is difficult to predict attenuation by aerosols. Although their concentration is closely related to the optical visibility, there is no single particle dimension distribution for a given visibility [28]. Due to the fact that the visibility is an easily obtainable parameter, either from airport or weather data, the scattering coefficient β_a can be expressed according to visibility and wavelength by the following expression [5]:

$$\beta_a = \left(\frac{3.91}{V}\right)\left(\frac{0.55\mu}{\lambda}\right)^i \tag{9}$$

Where:

V: is the visibility (Visual Range)$[km]$.

λ: is the incident laser beam wavelength$[\mu m]$.

i: is the size distribution of the scattering particles which typically varies from 0.7 to 1.6 corresponding to visibility conditions from poor to excellent.

Where:

$i = 1.6 \ for \ V > 50km.$

$i = 1.3 \ for \ 6 \ km \leq V \leq 50 \ km.$

$i = 0.585 \ V^{1/3} \ for \ V < 6 \ km.$

Since we are neglecting the absorption attenuation at wavelength of interest and Rayleigh scattering at terrestrial altitude and according to Eq. (2) and Eq. (5) then:

$$\beta_{scat} = \beta_a \tag{10}$$

The atmospheric attenuationτ is given as:

$$\tau = \exp(-\beta_a L) \tag{11}$$

The atmospheric attenuation in dB, τ can be calculated as follows:

$$\tau = 4.3429 \beta_a L \quad [dB] \tag{12}$$

3.5.3 RAIN

Rain is formed by water vapor contained in the atmosphere. It consists of water droplets whose form and number are variable in time and space. Their form depends on their size: they are considered as spheres until a radius of 1 mm and beyond that as oblate spheroids: flattened ellipsoids of revolution [5].

Rainfall effects on FSO systems

Scattering due to rainfall is called non-selective scattering, this is because the radius of raindrops (100 – 1000 μm) is significantly larger than the wavelength of typical FSO systems. The laser is able to pass through the raindrop particle, with less scattering effect occurring. The haze particles are very small and stay longer in the atmosphere, but the rain particles are very large and stay shorter in the atmosphere. This is the primary reason that attenuation via rain is less than haze. An interesting point to note is that RF wireless technologies that use frequencies above approximately 10 GHz are adversely impacted by rain and little impacted by fog. This is because of the closer match of RF wavelengths to the radius of raindrops, both being larger than the moisture droplets in fog [4]. The rain scattering coefficient can be calculated using Stroke Law see Eq. (13) [29]:

$$\beta_{rainscat} = \pi a^2 N_a Q_{scat}\left(\frac{a}{\lambda}\right) \tag{13}$$

Where:

a: is the radius of raindrop, (cm).

N_a: is the rain drop distribution, (cm^{-3}).

Q_{scat}: is the scattering efficiency.

The raindrop distribution N_a can be calculated using equation following:

$$N_a = \frac{R}{1.33(\pi a^3)V_a} \tag{14}$$

Where:

R: is the rainfall rate (cm/s),

V_a: is the limit speed precipitation.

Limiting speed of raindrop [29] is also given as:

$$V_a = \frac{2a^2 \rho g}{9\eta} \tag{15}$$

Where:

ρ: is water density, $(\rho = 1\ g/cm^3)$.

g: is gravitational constant, $g = 980\ cm/sec^2$.

η: is viscosity of air, $\eta = 1.8 * 10^{-4} g/cm.sec.$

The rain attenuation can be calculated by using Beer's law as:

$$\tau = \exp(-\beta_{rainscat}L) \tag{16}$$

3.6 Turbulence

Clear air turbulence phenomena affect the propagation of optical beam by both spatial and temporal random fluctuations of refractive index due to temperature, pressure, and wind variations along the optical propagation path [30][31]. Atmospheric turbulence primary causes phase shifts of the propagating optical signals resulting in distortions in the wave front. These distortions, referred to as optical aberrations, also cause intensity distortions, referred to as scintillation. Moisture, aerosols, temperature and pressure changes produce refractive index variations in the air by causing random variations in density [32]. These variations are referred to as eddies and have a lens effect on light passing through them. When a plane wave passes through these eddies, parts of it are refracted randomly causing a distorted wave front with the combined effects of variation of intensity across the wave front and warping of the isophase surface [33]. The refractive index can be described by the following relationship [34]:

$$n - 1 \approx 79 \times \frac{P}{T} \tag{17}$$

Where:

P: is the atmospheric pressure in [mbar].

T: is the temperature in Kelvin [K].

If the size of the turbulence eddies are larger than the beam diameter, the whole laser beam bends, as shown in Fig. 4. If the sizes of the turbulence eddies are smaller than the beam diameter and so the laser beam bends, they become distorted. Small variations in the arrival time of various components of the beam wave front produce constructive and destructive interference and result in temporal fluctuations in the laser beam intensity at the receiver.

3.6.1 Refractive index structure

Refractive index structure parameter C_n^2 is the most significant parameter that determines the turbulence strength. Clearly, C_n^2 depends on the geographical location, altitude, and time of day. Close to ground, there is the largest gradient of temperature associated with the largest values of atmospheric pressure (and air density). Therefore, one should expect larger values C_n^2 at sea level. As the altitude increases, the temperature gradient decreases and so the air density with the result of smaller values of C_n^2 [3].

In applications that envision a horizontal path even over a reasonably long distance, one can assume C_n^2 to be practically constant. Typical value of C_n^2 for a weak turbulence at ground level can be as little as $10^{-17} m^{-2/3}$, while for a strong turbulence it can be up to $10^{-13} m^{-2/3}$ or larger. However, a number of parametric models have been formulated to describe the C_n^2 profile and among those, one of the more used models is the Hufnagel-Valley [35] given by Eq. (18):

$$C_n^2(h) = 0.00594(v/27)^2 (10^{-5}h)^{10} exp(-h/1000) + 2.7\times 10^{-16}exp(-\frac{h}{1500})+A_o exp(-\frac{h}{100}) \tag{18}$$

Where:

h: is the altitude in [m].

v: is the wind speed at high altitude [m/s].

A_0: is the turbulence strength at the ground level, $A_o = 1.7 \times 10^{-14} m^{-2/3}$.

The most important variable in its change is the wind and altitude. Turbulence has three main effects [36]; scintillation, beam wander and beam spreading.

3.6.2 Scintillation

Scintillation may be the most noticeable one for FSO systems [9]. Light traveling through scintillation will experience intensity fluctuations, even over relatively short propagation paths. The scintillation index, σ_i^2 describes such intensity fluctuation as the normalized variance of the intensity fluctuations given by Eq. (19) [3]:

$$\sigma_i^2 = \frac{\langle (I-\langle I \rangle)^2 \rangle}{\langle I \rangle^2} = \frac{\langle I^2 \rangle}{\langle I \rangle^2} - 1 \tag{19}$$

Where:

$I = |E|^2$: is the signal irradiance (or intensity).

The strength of scintillation can be measured in terms of the variance of the beam amplitude or irradiance σ_i given by the following:

$$\sigma_i^2 = 1.23C_n^2 k^{7/6} L^{11/6} \tag{20}$$

Here, $k = 2\pi/\lambda$ is the wave number and this expression suggests that longer wavelengths experience a smaller variance.

Where the Eq. (20) is valid for the condition of weak turbulence mathematically corresponding to $\sigma_i^2 < 1$. Expressions of lognormal field amplitude variance depend on: the nature of the electromagnetic wave traveling in the turbulence and on the link geometry [3].

3.6.3 Beam spreading

Beam spreading describes the broadening of the beam size at a target beyond the expected limit due to diffraction as the beam propagates in the turbulent atmosphere. Here, we describe the case of beam spreading for a Gaussian beam, at a distance L from the source, when the turbulence is present. Then one can write the irradiance of the beam averaged in time as presented in Eq. (21) [36]:

$$I(l,r) = \frac{2P_o}{\pi \omega_{eff}^2(l)} exp\left(\frac{-2r^2}{\omega_{eff}^2(l)}\right) \tag{21}$$

Where:

P_o: is total beam power in W.

r: is the radial distance from the beam center.

The beam will experience a degradation in quality with a consequence that the average beam waist in time will be $\omega_{eff}(l) > \omega(l)$. To quantify the amount of beam spreading, describes the effective beam waist average as:

$$\omega_{eff}(l)^2 = \omega(l)^2(1+T) \tag{22}$$

Where:

$\omega(l)$: is the beam waist that after propagation distance L is given by:

$$\omega(l)^2 = \left[\omega_o^2 + \left(\frac{2L}{k\omega_o}\right)^2\right] \quad (m^2) \tag{23}$$

In which ω_o is the initial beam waist at $L = 0$, T: is the additional spreading of the beam caused by the turbulence. As seen in other turbulence figure of merits, T depends on the strength of turbulence and beam path. Particularly, T for horizontal path, one gets [37]:

$$T = 1.33\sigma_i^2 \Lambda^{5/6} \tag{24}$$

While the parameter Λ is given by:

$$\Lambda = \frac{2L}{k\omega^2(l)} \tag{25}$$

The effective waist, $\omega_{eff}(l)$, describes the variation of the beam irradiance averaged over long term.

As seen in other turbulence figure of merits, $\omega_{eff}(l)^2$ depends on the turbulence strength and beam path [37]. Evidently, due to the fact that $\omega_{eff}(l) > \omega(l)$ beamwill experience a loss that at beam center will be equal:

$$L_{BE} = 20 \, log_{10}\big(\omega(l)/\omega_{eff}(l)\big)(dB) \tag{26}$$

3.7 Total attenuation

Atmospheric attenuation of FSO system is typically dominated by haze, fog and is also dependent on rain. The total attenuation is a combination of atmospheric attenuation in the atmosphere and geometric loss.

Total attenuation for FSO system is actually very simple at a high level (leaving out optical efficiencies, detector noises, etc.). The total attenuation is given by the following [38]:

$$\frac{P_r}{P_t} = \frac{d_2^2}{(d_1 + (\theta L))^2} \times exp(-\beta L) \tag{27}$$

Where:

P_t: is the transmitted power [mw].

P_r: is the received power [mw].

θ: is the beam divergence [mrad].

β: is the total scattering coefficient [1/km].

Looking at this equation, the variables that can be controlled are the aperture size, the beam divergence and the link range. The scattering coefficient is uncontrollable in an outdoor environment. In real atmospheric situations, for availabilities at 99.9% or better, the system designer can choose to use huge transmitter laser powers, design large receiver apertures, design small transmitter apertures and employ small beam divergence. Another variable that can be controlled is link range, which must be of a short distance to ensure that the atmospheric attenuation is not the dominant term in the total attenuation [14].

3.8 Conclusion

FSO communication systems are affected by atmospheric attenuation that limits their performance and reliability. The atmospheric attenuation by fog, haze, rainfall, and scintillation has a harmful effect on FSO system. The majority of the scattering occurred to the laser beam is due to the Mie scattering. This scattering is due to the fog and haze aerosols existed at the atmosphere. This scattering is calculated through visibility. FSO attenuation at thick fog can reach values of hundreds dB. Thick fog reduces the visibility range to less than 50 m, and it can affect on the performance of FSO link for distances as small. The rain scattering (non-selective scattering) is wavelength independent and it does not introduce a significant attenuation in wireless IR links, it affect mainly on microwave and radio systems that transmit energy at longer wavelengths.

There are three effects on turbulence: scintillation, laser beam spreading and laser beam wander. Scintillation is due to variation in the refractive index structure of air, so if the light travelling through scintillation, it will experience intensity fluctuations. The Geometric loss depends on FSO components design such as beam divergence, aperture diameter of both transmitter and receiver. The total attenuation depends on atmospheric attenuation and Geometric loss. In order to reduce total attenuation, FSO system must be designed so that the effect of geometric loss and atmospheric attenuation is small.

4. Simulation results and analysis

FSO system used the laser beam to transfer data through atmosphere. The bad atmospheric conditions have harmful effects on the transmission performance of FSO. These effects could result in a transmission with insufficient quality and failure in communication. So, the implementation of the FSO requires the study of the local weather conditions patterns. Studying of the local weather conditions patterns help us to determine the atmospheric attenuation effects on FSO communication that occurs to laser beam at this area. In this part of this work, we shall discuss the effects of atmospheric attenuation, scattering coefficient during rainy and hazy days and atmospheric turbulence during clear days on the FSO system performance. Finally, we will calculate the atmospheric turbulence.

In this part we refer to the discussion and analysis of the effects of atmospheric attenuation, scattering coefficient and atmospheric turbulence on FSO system at weather conditions in the Republic of Yemen.

Results analysis is based on weather conditions data in Yemen which has been obtained from the (CAMA) for visibility are listed in Table (4) for the year 2008 and for wind velocity are listed in Table (10) for year 2003. The data of rainfall rate obtained from the (PAWR) are listed in Table (4) for the year 2008. Real data has been included in this analysis. Data has been collected and classified into two types:

1. Hazy days' data: which has been classified based on low and average visibility.
2. Rainy days' data: has been classified based on the rainfall rate at heavy, moderate and light rainfall.
3. Clear days' data: has been classified based on wind velocity.

Based on the above classification we have calculated the following:

1. Scattering coefficient at hazy and rainy days.
2. Atmospheric and total attenuation at hazy and rainy days and link range.
3. Effects of turbulence based on the wind velocity.

		Month	Jan.	Feb.	Mar.	Apr.	May	Jun.	Jul.	Aug.	Sep.	Oct.	Nov.	Dec.
Visibility in km	Sana'a	Av.	9.9	8	9.1	9.1	9.5	7.3	5.6	8.6	9.8	9.8	10	10
		Low	5	0.3	1	2	4	0.7	0.5	2	3	2	7	6
	Aden	Av.	9.7	8.3	9.1	9.2	9.4	7	5.7	8	9.1	9.2	9.8	9.7
		Low	6	2	5	7	3	3	1	5	3	0.05	8	7
	Taiz	Av.	9	8.4	9.7	9.7	9.9	8.9	8	8.8	9.9	9.5	9.3	8.9
		Low	0.05	0.05	4	4	0.05	2	3	1.5	1	1.5	0.1	0.1

Table 4. The Data of Visibility (km) obtained from CAMA for Year 2008.

4.1 Scattering coefficient in hazy days

In this part we shall discuss and analyze the effects of scattering coefficient on the FSO system performance during hazy days for Yemen and cities of Sana'a, Aden and Taiz. We will discuss and analyze the scattering coefficient during hazy days at low and average visibility. We will calculate the values of scattering coefficient using the Eq. 4.9 assuming that the size of distribution of scattering particles for low visibility is i = $0.585V^{1/3}$ and for average visibility is i = 1.3. The range of low visibility extends from 0.8 km to 5 km and the range of average visibility extends from 6.4 km to 5 km as shown in the Table (8).

Figure (5) represents the performance of scattering coefficient versus low visibility at wavelengths 780 nm, 850 nm and 1550 nm. This figure shows that the scattering coefficient inversely proportions with visibility. Scattering coefficient at low visibility 0.8 km is 4 km⁻¹, 3.9 km⁻¹ and 2.8 km⁻¹ for wavelengths 780 nm, 850 nm and 1550 nm respectively. The scattering coefficient of 5 km low visibility is 0.55 km⁻¹, 0.51 km⁻¹ and 0.28 km⁻¹ for wavelengths 780 nm, 850 nm, and 1550 nm respectively.

Figure (6) illustrates the scattering coefficient performance versus average visibility. When visibility is 6.4 km, scattering coefficient obtained was 0.39 km⁻¹, 0.35 km⁻¹ and 0.16 km⁻¹ for wavelengths 780 nm, 850 nm and 1550 nm respectively while at 9.7 km visibility it was 0.26 km⁻¹, 0.23 km⁻¹ and 0.11 km⁻¹ for wavelengths 780 nm, 850 nm, and 1550 nm respectively.

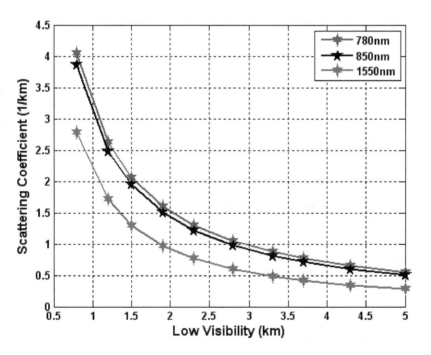

Fig. 5. Scattering Coefficient (km⁻¹) versus Low Visibility (km).

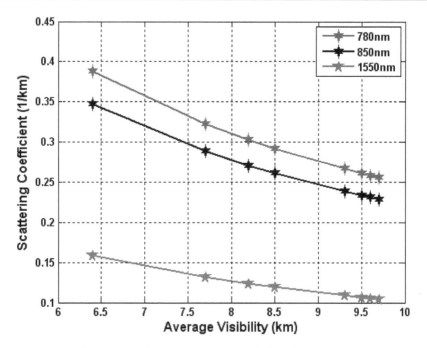

Fig. 6. Scattering Coefficient (km⁻¹) versus Average Visibility (km).

These results show that the wavelength 1550 nm is scattered less than wavelengths 850 nm and 780 nm. The scattering affects are less in average visibility compared with low visibility. This is because the distribution of particles density at low visibility is higher than the density of particles at average visibility. The results of scattering coefficient due to hazy days are given in the Table (5).

Visibility	Wavelength	From Scattering (km⁻¹)	To Scattering (km⁻¹)
Low	780 nm	4	0.55
	850 nm	3.9	0.51
	1550 nm	2.8	0.28
Average	780 nm	0.39	0.26
	850 nm	0.35	0.23
	1550 nm	0.16	0.11

Table 5. The Results of Scattering Coefficient due to Hazy Days.

4.2 Atmospheric attenuation in hazy days

In this part we will discuss and analyze the effects of atmospheric attenuation on the FSO system performance during hazy days. We have obtained the value of atmospheric attenuation from Eq. (12) assuming that the distance between transmitter and receiver is 1 km.

Figure (7) shows that the atmospheric attenuation inversely proportions with visibility. When the visibility is higher the effect of atmospheric attenuation is higher too. At low visibility of 0.8 km, atmospheric attenuation is 17.6 dB, 16.8 dB and 12.1 dB for wavelengths 780 nm, 850 nm and 1550 nm respectively. For 5 km low visibility, the atmospheric attenuation is about2.4 dB, 2.2 dB and 1.2 dB for wavelengths 780 nm, 850 nm and 1550 nm respectively.

Figure (8) represents the atmospheric attenuation versus average visibility. When visibility is 6.4 km, atmospheric attenuation is 1.7 dB, 1.5 dB and 0.69 dB for wavelengths 780 nm, 850 nm and 1550 nm respectively. For a 9.7 km visibility, the atmospheric attenuation is 1.1 dB, 0.99 dB and 0.46 dB for wavelengths 780 nm, 850 nm and 1550 nm respectively.

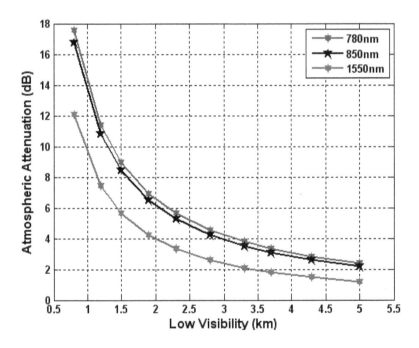

Fig. 7. Atmospheric Attenuation (dB) versus Low Visibility (km).

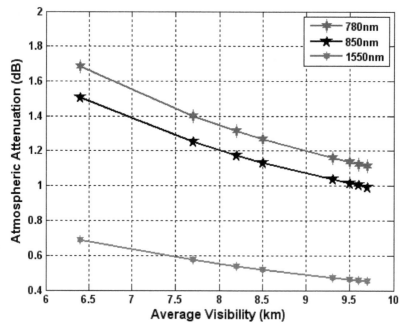

Fig. 8. Atmospheric Attenuation (dB) versus Average Visibility (km).

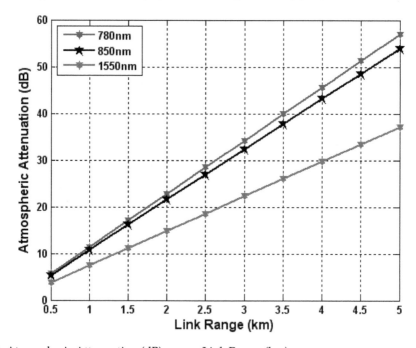

Fig. 9. Atmospheric Attenuation (dB) versus Link Range (km).

Figure (9) indicates to the atmospheric attenuation versus link range that extends from 0.5 km to 5 km. Here we assume that the visibility is 1.2 km and i = 0.585 * v$^{1/3}$. The more the distance between the transmitter and the receiver, the more the atmospheric attenuation is. This means that when the distance between the transmitter and the receiver increases, it is able to reduce the quality of transmission and effectiveness of FSO system. Atmospheric attenuation for link range 0.5 km at low visibility is 5.7 dB, 4.5 dB, 3.7 dB for wavelengths 780 nm, 850 nm and 1550 nm respectively. When the link range was about 5 km, atmospheric attenuation was 56.9 dB, 54 dB and 37.2 dB, for wavelengths 780 nm, 850 nm and 1550 nm respectively.

These results show that the attenuation at low visibility is higher than attenuation at average visibility. In addition, these readings have proved that the wavelength 1550 nm is capable to reduce the effect of atmospheric attenuation on FSO system. The distance between transmitter and receiver at low visibility should be reduced to avoid the effect of atmospheric attenuation on FSO system and improve its performance. The results of atmospheric attenuation due to hazy days are given in the Table (6).

Visibility	Wavelength	From	To
		Attenuation (dB)	Attenuation (dB)
Low	780 nm	17.6	2.4
	850 nm	16.8	2.2
	1550 nm	12.1	1.2
Average	780 nm	1.7	1.1
	850 nm	1.5	0.99
	1550 nm	0.69	0.46

Table 6. The Results of Atmospheric Attenuation due to Hazy Days.

4.3 Scattering coefficeint in rainy days

Figures below were plotted based on Eq. (14) assuming the water density ($\rho = 0.001$ g/ mm^3), gravitational constant (g = 127008 $*$ 10^{6}mm/hr^2), viscosity of air ($\eta = 0.0648$ g/ mm. hr) and scattering efficiency(Q = 2). The data of rainfall rate which listed in the Table (7) are divided into three states: light, moderate and heavy rain.

Month	Jan.	Feb.	Mar.	Apr.	May	Jun.	Jul.	Aug.	Sep.	Oct.	Nov.	Dec.
Sana'a	2.7	2.6	2.1	1	0	0	2.3	3	0	0	1	0.6
Aden	0	0	0	1	0.5	0	0	1.2	0	0	0	0
Taiz	2.4	0.5	3.75	5.77	5.41	3.1	5	4.72	4.02	2.4	1.5	0.5

(Rainfall rate (mm/hr))

Table 7. The Data of Rainfall Rate (mm/hr) obtained from (PAWR) for Year 2008.

Figure (10) illustrates the performance of scattering coefficients versus rainfall rate at light, moderate and heavy rain. The curves plotted were based on Eq. (14) assuming the radius of raindrop $a = 0.5$ mm. The scattering coefficient is proportional with rainfall rate, which showed that when the rainfall rate increases, the scattering coefficient increases too. For light rain the scattering coefficient is about 0.008 km^{-1} to 0.04 km^{-1}, about 0.055 km^{-1} to 0.086 km^{-1} for moderate rain and 0.10 km^{-1} to 0.16 km^{-1} for heavy rain. The highest scattering coefficient is about 0.16 km^{-1} in heavy rain. The impact of scattering on transmission of FSO system is more pronounced during heavy rainfall compared to moderate and light rainfall.

Figure (11) shows that the scattering coefficient versus raindrop radius. This figure illustrated that the radius of raindrop was important in evaluating the scattering effect. The radii of raindrop fall in the range of 0.1 mm to 0.8 mm. The scattering coefficient of the rain is independent of wavelength because the radii of rain particles are much bigger than laser wavelengths.

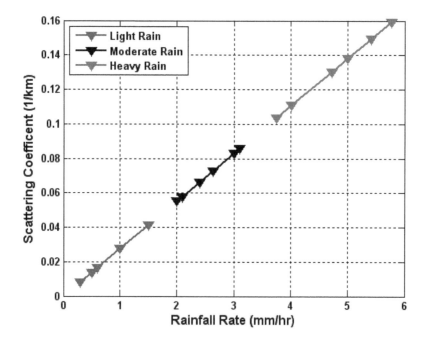

Fig. 10. Scattering Coefficient (km^{-1}) versus Rainfall Rate (mm/hr).

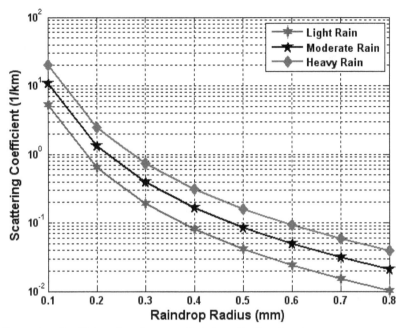

Fig. 11. Scattering Coefficient (km⁻¹) versus Raindrop Radius (mm).

The results of scattering coefficient due to rainy days are given in the Table (8).

Rainfall rate	From	To
	Scattering (km⁻¹)	Scattering (km⁻¹)
Light rain	0.0083	0.041
Moderate rain	0.055	0.086
Heavy rain	0.10	0.16

Table 8. The Results of Scattering Coefficient due to Rainy Days.

4.4 Atmospheric attenuation in rainy days

In this part we will discuss the effects of atmospheric attenuation on the performance of FSO system during rainy days. The effects of atmospheric attenuation on FSO systems during rainy days depended on rainfall rate intensity and raindrop radius.

Figure (12) shows the atmospheric attenuation versus rainfall rate. The curves plotted were based on Eq. 17 at light, moderate and heavy rain, assuming the radius of rain a = 0. 5 mm and transmission range L = 1 km. When the rainfall rate increases the effect of atmospheric attenuation on the FSO system increases too. Therefore influence of attenuation on transmission of FSO systems is more prominent during heavy rainfall compared to moderate and light rainfall. The atmospheric attenuation is about 0.036 dB to 0.18 dB for light rain, bout 0.24 dB to 0.37 dB for moderate rain and 0.45 dB to 0.69 dB for heavy rain. The highest attenuation is about 0.69 dB in heavy rain.

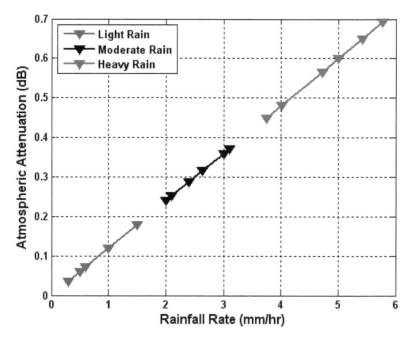

Fig. 12. Atmospheric Attenuation (dB) versus Rainfall Rate (mm/hr).

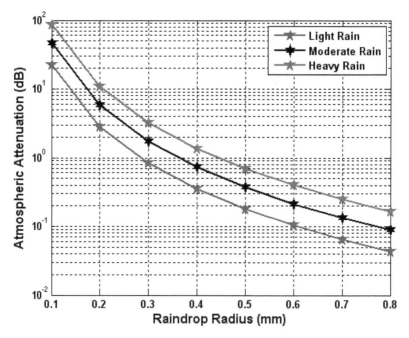

Fig. 13. Atmospheric Attenuation (dB) versus Raindrop Radius (mm).

Figure (13) illustrates that the atmospheric attenuation versus raindrop radius. The radius of rain particles falls in the range of 0.1 mm to 0.8 mm. This figure shows that the atmospheric attenuation decreases `when the radius of raindrop increases.

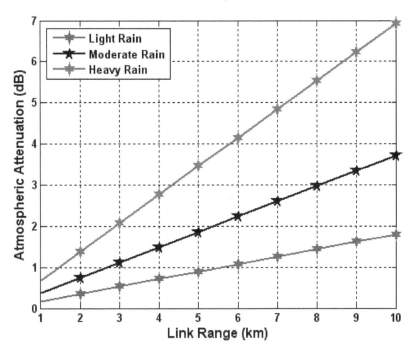

Fig. 14. Atmospheric Attenuation (dB) versus Link Range (km).

Figure (14) indicates the atmospheric attenuation versus link range. This figure was plotted based on Eq. (17) assuming the raindrop radius is 0.5 mm. For 0.5 km link range the atmospheric attenuation is about 0.18 dB for light rain, 0.37 dB for moderate rain and 0.69 dB for heavy rain. For 10 km link range the atmospheric attenuation is about 1.8 dB for light rain, 3.7 dB for moderate rain and 6.9 dB for heavy rain. The atmospheric attenuation results due to rainy days are given in the Table (9).

Rainfall rate	Atmospheric Attenuation (dB)	
	From	To
Light rain	0.036	0.18
Moderate rain	0.24	0.37
Heavy rain	0.45	0.69

Table 9. The Results of Atmospheric Attenuation due to Rainy Days.

4.5 Atmospheric turbulence

The purpose here is to discuss the relationship for calculating irradiance variance, beam spreading and loss beam center for a range of parameters. We used the wavelengths of 780 nm, 850 nm & 1550 nm.

Month	Jan.	Feb.	Mar.	Apr.	May	Jun.	Jul.	Aug.	Sep.	Oct.	Nov.	Dec.
Sana'a	15. 6	17.4	15.2	15.2	14.8	16.1	17.6	16.5	18.9	16.9	16.3	14.8
Aden	20.6	18.5	20.9	20	13.3	18.3	21.3	20.6	14.6	16. 7	20	19.1
Taiz	12.4	14.6	16.1	17.2	17	18.3	21. 7	17	17	14.8	14.4	13

(Wind Velocity (km\hr))

Table 10. The Data of Wind Velocity (km/hr) obtained from (CAMA) for Year 2003.

Figure (15) illustrates the log irradiance variance versus the link range for 780 nm, 850 nm and 1550 nm wavelengths. This figure was plotted based on Eq. 21. As the link range increases the variance (scintillation) increases too. For a 0.5 km link range, the variance is about 0.087, 0.079and 0.039 for wavelengths 780 nm, 850 nm and 1550 nm respectively. For a 5 kmlink range, the variance is about 5.9, 5.4 and 2.7 for 780 nm, 850nm and 1550 nm respectively. These results show that the use of a wavelength of 1550 nm is able to reduce the variance "atmospheric turbulence" effect on the FSO systems.

Figure (16) indicates the comparison between the beam spreading on a distance (L) from the transmitter, in case the atmospheric turbulences and its absence. Figure (15) was plotted based on Eq. 23 assuming the spot size of the beam at the transmitter (with the distance L = 0) equals 8 mm. At the distance 0.5 km from transmitter, the spot size of the beam is $\omega(l)$ = 0.032 m in case of absence turbulence and $\omega_{eff}(l)$ = 0.032 m in case of turbulences. At the distance 5 km, the $\omega(l)$= 0.31 m and $\omega_{eff}(l)$ = 0.33m. From the above results, we conclude the expansion of the spot size of the beam depends on the distance between transmitter and receiver and on the atmospheric turbulence on the along of transmission range.

The loss beam at center (dB) depends on transmission range and wavelength as shown Fig. (17). The loss beam at the center increases, corresponding to the increase of Link range. At the distance 0.5 km, the loss beam at center = 0.0454 dB, 0.0383 dB and 0.0116 dB for wavelengths 780 nm, 850 nm & 1550 nm respectively. At the distance 5 km, the loss beam at center is 2.4 dB, 2.1 dB and 0.72 dB for wavelengths 780 nm, 850 nm& 1550nm respectively.

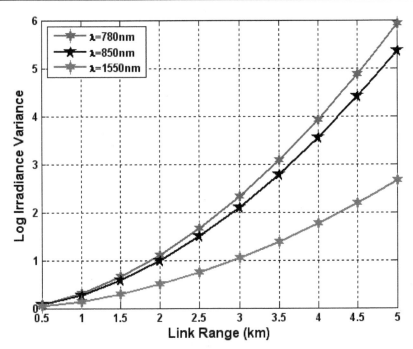

Fig. 15. Log Irradiance Variance Scintillation versus Link Range (km).

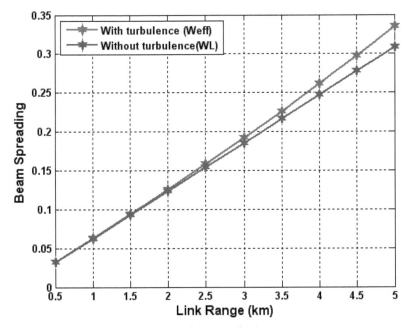

Fig. 16. Beam Spreading (m) versus the Link Range (km).

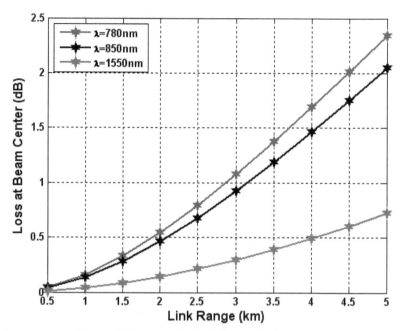

Fig. 17. Loss at Beam Center (dB) versus the Link Range (km).

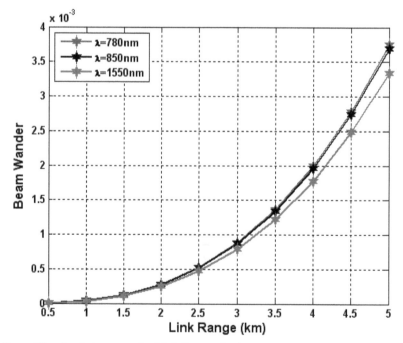

Fig. 18. Beam Wander (m) versus the Link Range (km).

Figure (18) indicates to the beam wander versus link range. The beam wanders increases corresponding to increasing in the link range. At 0.5 km transmission range, the beam wander is 0 m for 780 nm, 850 nm and 1550nm wavelengths respectively and at 5 kmlink range the beam wander is 0.0037 m, 0.0037 m, and 0.0033 m for 780 nm, 850nm and 1550 nm wavelengths respectively. From the above results, we conclude that the loss beam at center for 780 nm and 850 nm wavelengths is more than the loss at 1550nm wavelength. So to reduce the loss beam at center we suggest to reduce the link range and 1550 nm wavelength must be used. The results of atmospheric turbulence effect due to clear days are given in the Table (11).

Link Range (km)	Wavelength	Scintillation $(m^{-3/2})$	Loss at Beam center (dB)	Beam wander (m)	W (L) (m)	W_{eff} (L) (m)
0.5 km	780 nm	0.087	0.045	0.0		
	850 nm	0.076	0.038	0.0	0.032	0.032
	1550 nm	0.039	0.012	0.0		
5 km	780 nm	5.92	2.35	0.0037		
	850 nm	5.35	2.050	0.0037	0.31	0.33
	1550 nm	2.66	0.73	0.0033		

Table 11. The Results of Atmospheric Turbulence due to Clear Days.

4.6 Conclusion

In this chapter, we focused on haze, rain and turbulence effects on FSO systems. Mie scattering occurs in hazy days and it depends on wavelength. The scattering coefficient on hazy days is determined by using Beer's Law. From the results analysis and data in the Table 5.1 the fog and haze represent the most important atmospheric scatters. Their attenuation, which can reach about 17.6 dB at 1.8 km low visibility in Yemen and 163.5 dB (corresponding to very thick fog), at 0.05 km low visibility is in Taiz city. This attenuation value affects the performance of a FSO link for distances as small. Wavelength 1550 nm is less scattered from the wavelengths 850 nm & 780 nm and it is not harmful to the human eyes.

Rain does not introduce a significant attenuation in FSO systems links in Yemen. This is due to the rainfall affect mainly radio and microwave systems that use a longer wavelengths and attenuation at heavy rain 5.77 mm/hr in Yemen about 0.69 dB, is very small compared with attenuation due to fog. Therefore the effect of rain is neglected in Yemen. Atmospheric turbulence will change in refractive index structure of air from one area to another. Atmospheric turbulence fluctuates intensity of the laser beam. Scintillation is wavelength and distance dependent. We can reduce the effect of the turbulence by enlarging the diameter of the receiver's aperture or setting tracking system at the receiver. The results indicate that the attenuation depends on weather conditions which are uncontrollable and transmission range which can be controlled; hence, it is considered an important element in the design of FSO system. So, to improve the performance of FSO system, we must reduce the transmission range and use wavelength 1550 nm.

5. General conclusion of this chapter

FSO system can spread as a reliable solution for high bandwidth and short distance. There are some factors which must be taken into consideration during the design of FSO system as controllable and uncontrollable factors. Controllable factors include wavelength, transmission range, beam divergence, loss occurred between transmitter and receiver and detector sensitivity. Uncontrollable factors include visibility, rainfall rate, raindrop radius, atmospheric attenuation and scintillation.

Atmospheric attenuation may be absorption or scattering. Absorption lines at the visible and IR wavelengths are narrow and separated. So, we can ignore absorption effect at the wavelength identified as atmospheric windows. Wavelength at FSO system must be eye safe and able to transmit a sufficient power during the bad weather condition. Mie scattering represents the main affects on FSO systems. The main cause of Mie scattering is fog and hazy. Attenuation caused by fog in Yemen is so important for Taiz as the low visibility range can less than 0.05 km during the extensive fog according to the data taken from metrology authority. Transmission in this city may be cut off, so the distance between the transmitter and receiver must be reduced. However, Sana'a and Aden cities the weather is clear during the whole year in comparison with Taiz city. Rayleigh scattering we can ignore it at the visible and infrared wavelength as its effect on the ultraviolet wavelengths is huge. This scattering occurs when the molecules size is less than the wave length of the laser beam.

Non-selective scattering independent on wavelength and occurs when the molecules size is bigger than wavelength and it occurs due to the rainfall. Generally, FSO system is so adequate in Yemeni environment according to the previous results. The performance of wavelength 1550 nm is better at the bad weather conditions in comparison with wavelengths 850 nm and 780 nm. Furthermore, the wavelength 1550 nm allows a high power may reach to over 50 times in comparison with the wavelengths 850 nm & 780 nm.

By analyzing results obtained at chapter four, we conclude that we are able to improve the performance of transmission of FSO system at the bad weather conditions by using the wavelength 1550 nm and short distance between transmitter and receiver.

6. References

[1] Weichel H., "Laser Beam Propagation in the Atmosphere", SPIE, Optical Engineering Press, Vol. TT-3, 1990.
[2] A. K. Majumdar, J. C. Ricklin, "Free Space Laser Communications Principles and Advances", Springer ISBN 978-0-387-28652-5, 2008.
[3] H. Hemmati, "Near-Earth Laser Communications", California, Taylor & Francis Group, Book, LLC, 2008.
[4] H. Willebrand and B. S. Ghuman, "Free-Space Optics Enabling Optical Connectivity in Today's Networks", SAMS, 0-672-32248-x, 2002.
[5] B. Olivieret, et al., "Free-Space Optics, Propagation and Communication", Book, ISTE, 2006.

[6] Roberto Ramirez-Iniguez, Sevia M. Idrus and Ziran Sun, "Optical Wireless Communications IR for Wireless Connectivity", Taylor & Francis Group, Book, CRC Press, 2007.

[7] N. Araki, H. Yashima, "A Channel Model of Optical Wireless Communications during Rainfall", IEEE, 0-7803-9206, 2005.

[8] Z. Ghassemlooy and W. O. Popoola , "Terrestrial Free Space Optical Communication" , Optical Communications Research Group, NCR Lab., Northumbia University, Newcastle upon Tyne, UK, 2010.

[9] Nadia B. M. Nawawi, "Wireless Local Area Network System Employing Free Space Optic Communication Link", A Bachelor Degree thesis, May 2009.

[10] Delower H. and Golam S. A., "Performance Evaluation of the Free Space Optical (FSO) Communication with the Effects of the atmospheric Turbulence", A Bachelor Degree thesis, January 2008.

[11] M. Zaatari, "Wireless Optical Communications Systems in Enterprise Networks", the Telecommunications Review, 2003.

[12] I. I. Kim, B. McArthur and E. Korevaar, "Comparison of laser beam propagation at 785 nm and 1550 nm in fog and haze for optical wireless communications", Optical Wireless Communications III, SPIE, 4214, 2001.

[13] Jeganathan, Muthu and Pavellonov, "Multi-Gigabits per Second Optical Wireless Communications", 2000.

[14] Bloom S. E. Korevaar, J. Schuster, H. Willebrand, "Understanding the Performance of Free Space Optics", Journal of Optical Networking, 2003.

[15] Alexander S. B., "Optical Communication Receiver Design", (ISOE) SPIE, 1997.

[16] D. Romain, M. Larkin, G. Ghayel, B. Paulson and g. Nykolak , "Optical wireless propagation, theory vs. Experiment", Optical Wireless Communication III, Proc. SPIE, Vol.4214, 2001.

[17] W. E. K. Middleton, "Vision through the Atmosphere", University of Toronto Press, Toronto, 1963.

[18] I. I. Kim, B. McArthur, and E. Korevaar, "Comparison of laser beam propagation at 785 nm and 1550 nm in fog and haze for optical wireless communications", Proc. SPIE, 4214, pp. 26-37, 2000.

[19] S. G. Narasimhan and S. K. Nayar, "Vision and the Atmosphere", 2007.

[20] B. Naimullah, S. Hitam, N. Shah, M. Othman and S. Anas, "Analysis of the Effect of Haze on Free Space Optical Communication in the Malaysian Environment", IEEE, 2007.

[21] F. T. Arecchi, E. O. Schulz-Dubois, "Laser Handbook". North-Holland publishing Co; 3 Reprint edition (Dec 1972)

[22] Willebrand, Heinz A., Ghuman B. S. "Fiber Optic Without Fiber Light Pointe Communications", 2001.

[23] N. J Veck, "Atmospheric Transmission and Natural Illumination (visible to microwave regions) ", GEC Journal of Research, 3(4), pp. 209 – 223, 1985.

[24] M. A. Bramson, "in Infrared Radiation", A handbook for Applications, Plenum Press, p. 602, 1969.

[25] Earl J. McCartney, "Optics of the Atmosphere: Scattering by Molecules and Particles", Wiley & Sons, New York, 1997.

[26] B. Bova, S. Rudnicki, "The Story of Light", Source book ISBN, 2001.

[27] P. P. Smyth et. al., "Optical Wireless Local Area Networks Enabling Technologies", BT Technology Journal, 11(2), pp. 56–64, 1993.

[28] M. S. Awan, L. C. Horwath, S. S. Muhammad, E. Leitgeb, F. Nadeem, M. S. Khan, "Characterization of Fog and Snow Attenuations for Free-Space Optical Propagation", Journal, Vol. 4, No. 8, 2009

[29] Achour M., "Simulating Atmospheric Free-Space Optical Propagation part I, Haze, Fog and Low Clouds, Rainfall Attenuation", Optical Wireless Communications, Proceedings of SPIE, 2002.

[30] I. Kim, R. et. al. , "Wireless Optical Transmission of Fast Ethernet, FDDI, ATM, and ESCON Protocol Data using the Terra Link Laser Communication System" , Opt. Eng. , 37 , pp. 3143-3155 , 1998.

[31] J. Li, and M. Uysal, "Achievable Information Rate for Outdoor Free Space Optical", Global Telecommunications Conference, Vol.5, pp .2654-2658, 2003.

[32] Gary D. Wilkins, "The Diffraction Limited Aperture of Atmosphere and ITS Effects on Free Space Laser Communication".

[33] Kim Issac I. And Eric Korevaar, "Availability of Free Space Optics (FSO) and Hybrid FSO/RF Systems" Optical Access, Incorporated, 2002.

[34] Fried, D. L. "Limiting Resolution Down Through the Atmosphere", J Opt. SOC. AM Vol., 56 No 10, 1966

[35] D. S. Kim, "Hybrid Free Space and Radio Frequency Switching", Master Degree Thesis, University of Maryland, 2008.

[36] Tyson, R. K., "Introduction to Adaptive Optics", SPIE Press, 2000.

[37] L. C. Andrews and R. L. Phillips, "Laser Beam Propagation through Random Media", SPIE Optical Engineering , 1998.

[38] Brown, Derek "Terminal Velocity and the Collision/Coalescence Process", 2003.

Part 2

Optical Communications Systems:
Amplifiers and Networks

4

Hybrid Fiber Amplifier

Inderpreet Kaur[1] and Neena Gupta[2]
[1]Rayat and Bahra Institute of Engineering, Mohali,
[2]PEC University of Technology (Formally Punjab Engineering College), Chandigarh
India

1. Introduction

The advent of telecommunications in 1870s completely revolutionized the world of communications. Metallic cables consisting of twisted wire cables, co-axial cables were the media of choice for many years. These could be used efficiently up to frequencies of 10MHz but the system performance degraded beyond this range. However, with the increasing demand for telephone services, it was necessary to find an alternative medium for telephony to cope up with the high demand. The development of low loss optical fibers gave a solution to this problem and their use revolutionized the speed of telecommunication. Optical fibers have become an unavoidable part of any high speed communication system due to its high information carrying capacity, high bandwidth and extremely low loss. The transmission performance of the optical communication systems is limited by various effects such as attenuation, dispersion, non- linearity, scattering etc, which degrade the level of the signal. To compensate for all these limitations the signals have to be regenerated within the transmission link after some distance. While setting up the transmission link, it is to be ensured that the signal can be retrieved intelligibly at the receiving end. This can be done either by using optoelectronic repeaters or optical amplifiers. In optoelectronic repeaters the optical signal is first converted into an electric signal, then amplified in electric domain and finally converted back into optical signals. Regeneration by making use of repeater is a traditional way to compensate for loss and degradation along the transmission medium. Such regenerators become quite complex and expensive for dense wavelength division multiplexed (DWDM) lightwave systems. This process works well for moderate speed single wavelength operation but it can be fairly complex and expensive for high- speed multi- wavelength systems. Moreover these so called opto-electronic repeaters once installed into the system can not be upgraded to higher bit rates. Thus a great deal of effort has been spent to develop all optical amplifiers. These devices operate in the optical domain to boost the power level of the signals. In the history of optical fiber communication systems, the advent of optical amplifier was an important milestone. Optical amplifiers can amplify the optical signals directly without requiring its conversion to the electric domain. The development of optical amplifiers started in early eighties and their use for long haul communication systems became widespread during late nineties. Optical amplifiers provided flexibility while upgrading the installed transmission links to higher bit rates. This flexibility of the bit rates allows overcoming the electrical bottleneck of an electric repeater, which was unable to transmit at high bit rates. The opto-electronic repeaters provided with maximum of 40-80 Gbps bit rate.

1.1 DWDM systems

To increase the transmission capacity of a single fiber, DWDM is used. DWDM is a technology, which combines large number of independent information carrying wavelengths onto the same fiber. A characteristic of DWDM is that the discrete wavelengths form an orthogonal set of carriers, which can be separated, routed and switched without interfering with each other. This isolation between channels holds as long as the total optical power intensity is kept sufficiently low to prevent non linear effects e.g. Stimulated Brillouin Scattering (SBS) and Four Wave Mixing processes (FWM) from degrading the link performance. The implementation of DWDM system requires a variety of passive and active devices to combine, distribute, isolate and amplify optical power at different wavelengths. Passive devices require no external control for their operation, so they are less flexible. The wavelength dependent performance of active devices can be controlled electronically, so they provide more flexibility to the network system. Optical amplifiers, tunable filters and tunable sources are integral part of any DWDM system. The key component of DWDM system is optical amplifier. In DWDM system, it is desirable to set a very narrow grid of optical carriers in order to allow more channels in the same optical bandwidth. This not only demands an optical amplifier with high gain but also very broad and flat gain profile to ensure a nearly identical amplification factor in every channel. Figure 1 shows the implementation of active as well as passive components in a typical DWDM system having post amplifier, in-line amplifier and preamplifier [Keiser 2009; Mynbaev 2003].

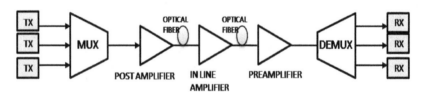

Fig. 1. Implementation of A DWDM System Having Various Types of Optical Amplifiers

2. Review of an optical amplifier

An optical amplifier works on the principle of stimulated emission. Optical amplifier increases the level of signal through this process. The mechanism for stimulated emission is same as that for lasers. The operation of laser diodes that are required for the fiber amplifier is similar to the external current injection method (which is used in semiconductor optical amplifiers, SOAs, discussed later). This method is the pumping method used to create population inversion needed for gain mechanism in fiber amplifiers. The sum of injection, stimulated emission and spontaneous recombination rates gives the rate equation that governs the carrier density N (t) in the excited state of both the amplifiers. This carrier density is given by equation (1) [Keiser 2009;Mynbaev 2003].

$$\frac{\partial N(t)}{\partial t} = R_1(t) - R_2(t) - \frac{N(t)}{\tau_r} \tag{1}$$

where,

$$R_1(t) = \frac{J(t)}{qd} \tag{2}$$

is the external pumping rate from the injection current density J(t) into an active layer having thickness d, τ_r is the combined time constant coming from carrier-recombination mechanisms and spontaneous emission, and

$$R_2(t) \cong gv_g N_p \tag{3}$$

is the net stimulated emission rate. Here, v_g is the group velocity of incident light, N_p is the photon density and g is the overall gain per unit length. The photon density N_p is dependent on optical signal power, energy of photons, group velocity and dimensions of active area of optical amplifier.

This photon density N_p is given by equation (4),

$$N_p = \frac{P_s}{(hv)(wd)v_g} \tag{4}$$

In equation (4), P_s is the signal power, v_g is group velocity, w and d are the width and thickness of active area of optical amplifier respectively. The difference between the structure of optical amplifiers and laser diodes is that there is no feedback system in optical amplifiers. So, for boosting an incoming signal optical amplifier requires a pump. The pump supplies energy to the electrons in an active medium, which in turn causes population inversion. An incoming signal photon triggers these excited electrons to drop to lower levels through a stimulated emission process, thereby producing an amplified signal. The amplifier is connected with the optical fiber through a fiber- to- amplifier coupler. The basic components of an optical amplifier are shown in the figure 2) [Keiser 2009;Mynbaev 2003].

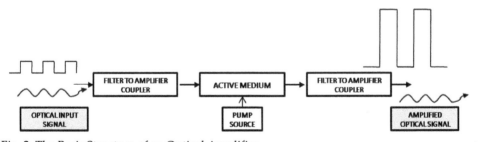

Fig. 2. The Basic Structure of an Optical Amplifier

The optical gain depends on the frequency/ wavelength of the signal. Let us consider a medium of two level systems for demonstrating the dependence of gain on frequency. The gain coefficient of such a medium can be written as below [Agarwal 2003]:

$$g(\omega) = \frac{g_0}{1 + (\omega - \omega_0)^2 T^2 + \dfrac{P}{P_{sat}}} \tag{5}$$

Where, g_0 is the peak value of the gain, ω is the optical frequency of the incident signal, ω_0 is the atomic transition frequency , P is the optical power of the signal being amplified, P_{sat} is the saturation power and T is the dipole relaxation time. In the unsaturated region, $P/P_s \ll 1$. So, the gain coefficient becomes

$$g(\omega) = \frac{g_0}{1+(\omega-\omega_0)^2 T^2} \qquad (6)$$

This equation shows that the gain reaches its maximum when the incident frequency coincides with the atomic transition frequency. Another term associated with optical amplifiers is amplification factor or amplifier gain (G) defined as:-

$$G = \frac{P_{out}}{P_{in}} \qquad (7)$$

Where P_{in} and P_{out} are the input and output powers of the continuous wave signal being amplified.

3. Applications of optical amplifiers

Optical amplifiers have found many applications ranging from ultra long undersea links to short links in access networks[Keiser 2009; Mynbaev 2003; Olsson1989 & Agarwal 2003].

- In-line amplifier
- Pre-amplifier
- Post -amplifier

In-line amplifier: This is used as a repeater along the link at intermediate points. It can be used to compensate for transmission loss and increase the distance between regenerative repeaters, as shown in figure 3a.

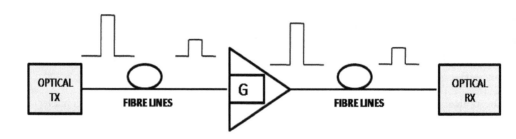

Fig. 3a. Optical Amplifier as In-Line Amplifier

Pre-amplifier: This is used before the photo detector at the receiver in order to strengthen the weak received signal. This increases the sensitivity of the detector effectively. This configuration is shown in figure 3b.

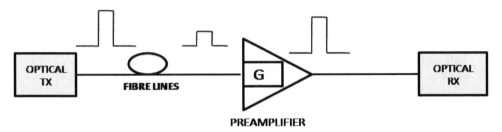

Fig. 3b. Optical Amplifier as Preamplifier

Post -amplifier: This is used at the transmitting end, after the source and operates near the saturation region. The power launched into the fiber is enhanced and so the repeater span can become large. This serves to increase the transmission distance by 10-100km depending on the amplifier gain and fiber loss. This configuration is shown in figure 3c.

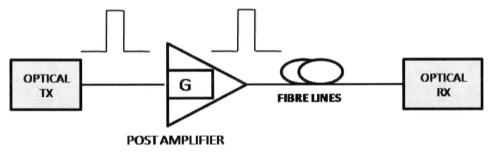

Fig. 3c. Optical Amplifier as Post Amplifier

4. Types of optical amplifiers

The optical amplifiers which find widespread use in communication systems can be classified into three categories:-

1. Fiber Raman Amplifier (FRA)
2. Erbium Doped Fiber Amplifier (EDFA)
3. Semiconductor Optical Amplifier (SOA)

The first two types, Fiber Raman Amplifier (FRA) and Erbium Doped Fiber Amplifier (EDFA) can be efficiently coupled to the transmission fiber by splicing with a minimum coupling loss. Of these two, EDFA requires lesser power for the pump source and the pump power requirements can be easily met by semiconductor laser diodes. Besides, the gain characteristics of EDFA are insensitive to polarization. Semiconductor Optical Amplifier (SOA) has the advantages of smaller size and lower power consumption. Its dimensional compatibility with the transmission fiber is obviously not as good as the fiber amplifier. However, SOA is suitable for optoelectronic integrated circuits.Table1 shows the basic difference between the three optical amplifiers and Table 2 shows the comparison of optical amplifiers.

Type of optical amplifier	Material required	Operating Working band
Semiconductor Optical Amplifier (SOA)	Semiconductor material from group III and V. e.g.phosphorous, gallium, indium and arsenic	O-Band and C-Band
Erbium Doped Fiber Amplifier (EDFA)	Lightly doping silica or tellurite with rare earth element i.e. erbium.	O-Band, S-Band, C-Band and L-Band
Fiber Raman Amplifier (FRA)	Raman Lasers	All Operating Bands

Table 1. Difference of materials and operating bandwidth of three optical amplifiers

S. No.	Parameter	Semiconductor Optical Amplifier (SOA)	Erbium Doped Fiber Amplifier (EDFA)	Fiber Raman Amplifier (FRA)
1	Gain(dB)	>30	>40	>25
2	Bandwidth (3dB)	60	30-60	Pump dependent
3	Max. Saturation (dBm)	18	22	0.75 X pump
4	Noise Figure (dB)	8	5	5
5	Pump Power	<400mA	25dBm	>30dBm
6	Wavelength(nm)	1260-1650	1530-1560	1260-1650
7	Time Constant	2×10^{-9} s	10^{-2}s	10^{-15} s
8	Size	Compact	Rack Mounted	Bulk Module
9	Cost factor	Low	Medium	High
10	Polarization Sensitivity	Yes	No	No

Table 2. Comparison of Optical Amplifiers

Although gain bandwidth of semiconductor laser amplifiers is ideally large, they have several drawbacks like polarization sensitivity, interchannel cross- talk and large coupling losses. Fiber amplifiers are preferable since the coupling loss due to fusion splice is negligible for them. Fiber amplifiers are also insensitive to polarization and have negligible noise for interchannel cross talk, which is one of the main noise sources in multichannel transmission or Dense Wavelength Division Multiplexing (DWDM). These reasons and available gain properties make the fiber amplifiers very suitable for modern optical transmission.

4.1 Advantages of EDFA

It is clear that EDFAs are the best choice for optical amplification in present lightwave systems. Erbium (Er: 68) is used as dopant into glass host (fiber) and the 'doped fiber' is used as an amplifying medium. Er-doped fibers give an amplified output around 1550nm [Desurvire 2002; Becker, Olsson & Simpson 1999; Sun et.al.1997]. The EDFA is one of the key devices used for dense wavelength division multiplexed (DWDM) transmission systems. EDFAs are revolutionizing lightwave systems by reducing system costs and enhancing network performance. Some of the advantages offered by EDFAs are:

- High gain (~50dB)
- High output power (>100mW)
- Low noise figure (~4dB)
- Less gain variation
- Wide bandwidth of operating suiting DWDM
- Inherent compatibility to transmission fiber with low insertion loss
- Cross talk immunity in multichannel systems

5. Working principle of EDFA

The invention of the Erbium Doped Fiber Amplifier (EDFA) in the late eighties was one of the major events in the history of optical communication systems. It provided new life to the research of technologies that allow high bit rate transmission over long distances. EDFA has a narrow high gain peak at 1532nm and a broad peak with a lower centered at 1550nm. The use of an increasing number of channels in the present day DWDM optical networks requires a flat gain spectrum across the whole usable bandwidth. Owing to their versatility, useful gain bandwidth, high pumping efficiency and low intrinsic noise, EDFAs are the amplifier of choice for most of the network applications. They are based on single mode optical fibers with cores that have been doped, typically to a few hundred part per million, with the trivalent erbium ion, Er^{3+}. The gain is provided through stimulated emission, as in laser. The Er^{3+} ion acts mostly as a three level system, in which the main participants are the $^4I_{15/2}$ ground state, the $^4I_{13/2}$ first excited level and the $^4I_{11/2}$ second excited level. The energy level diagram of Er^{3+} is shown in figure 4) [Keiser 2009;Mynbaev 2003].

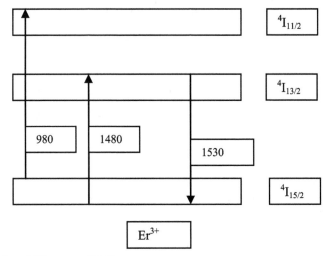

Fig. 4. Energy Level Diagram of Er3+

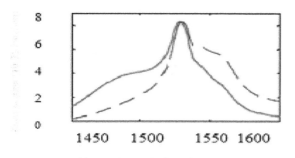

Fig. 5. Absorption and Emission Spectra of EDFA

EDFAs are of particular interest in telecommunications, because their emission spectrum shows a gain of more than 20dB over the range of 1530-1560nm. This is also the third window used in optical communication. The absorption spectrum reveals that good absorption takes place around 380nm, 520nm, 800nm, 980nm, and 1480nm. The absorption bands at shorter wavelengths are not of interest owing to the non- availability of semiconductor laser diodes at these wavelengths. At 980nm and 1480nm, efficient laser diodes are available and therefore used as pump sources.

6. Limitations of EDFA

The main practical limitation of an EDFA stems from the spectral non-uniformity of the amplifier gain. As a result, different channels of a DWDM system are amplified by different amounts. These problems become quite severe in long-haul systems, employing cascaded chain of EDFAs. Secondly, for many EDFA deployments, automatic gain control (AGC) is used to ensure that the output signal power is proportional to the input power. However, there are times when a constant optical signal output, independent of input power, is more desirable, e.g., in an optical preamplifier at an optical receiver [Qiao & Vella 2007]. The figure 6) [Keiser 2009;Mynbaev 2003].Figure 6 shows the gain spectrum of EDFA, from which it is clear that EDFA has peak gain at 1530nm, beyond which the gain reduces slightly and remains flat almost until 1550nm. After that, the gain reduces sharply. Several gain flattening techniques of EDFA are available [Lee et,al 1996; Ono et.al.1997; Kim et.al.1998; Park et.al.1998; Kawai et.al.1999; Yun et al. 1999; Lu & Chu 2000; Pasquale & Federighi 1995; Kemtchou et. al.1996; Hwang et al.2000; .Bakshi et.al.2001; Sohn et.al.2002; Arbore et.al.2002; Kaur & Gupta 2009 and Lobo et.al. 2003]

So, EDFAs are widely used in the C-band (1530-1560nm) for optical communication networks. So, there is a necessity to improve the amplification bandwidth of EDFA (i.e. broadening as well as flattening of gain spectrum). This would help to cater the needs of present day communication systems. In order to overcome this limitation of EDFA, different doping elements are coming into existence. One of such doping material is thulium and the doped fiber amplifier is known as Thulium Doped Fiber Amplifier (TDFA). TDFAs are highly viable alternative to meet out the limitations of EDFAs and have bright future prospects to be used in optical communication systems.

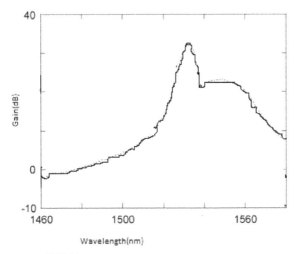

Fig. 6. Gain Spectrum of EDFA

7. Role of TDFA in communication systems

The optical fiber can be doped with any of the rare earth element, such as Erbium (Er), Ytterbium (Yb), Neodymium (Nd) or Praseodymium (Pr), Thulium (Tm). The host fiber material can be either standard silica, a fluoride based glass or a multicomponent glass. The operating regions of these devices depend on the host material and the doping elements. Fluorozirconate glasses doped with Pr or Nd are used for operation in the 1300nm window, since neither of the ions can amplify 1300nm signals when embedded in silica glass. The next popular material for long haul telecommunication applications is a silica fiber doped with Thulium, which is known as Thulium Doped Fiber Amplifier (TDFA). In some cases as Yb is added to increase the pumping efficiency and the amplifier gain. The TDFA are used in S-band (1460-1530nm). The energy state diagram of Tm^{3+} is shown in figure 7 [Aozasa et.al.2008].

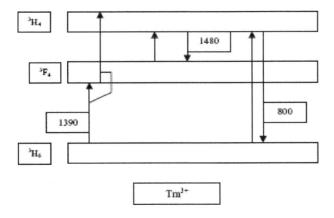

Fig. 7. Energy Level Diagram of Tm^{3+}

Tm^{+3} has three energy levels that are considered with respect to the Tm^{+3} populations. TDFA uses upconversion pumping method. The upconversion pumping consists of the two-step excitation of $^3H_6$ to $^3F_4$ and $^3F_4$ to $^3H_4$ with the same pump wavelength and this makes it possible to form a population inversion state between $^3F_4$and $^3H_4$. The gain and loss of TDFA in the 1460-1530nm wavelength region are determined not only by the excited state absorption (ESA) ($^3F_4$ to $^3H_4$) and stimulated emission(SE) ($^3H_4$ to $^3F_4$) but also by the ground state absorption (GSA) ($^3H_6$ to $^3F_4$) The absorption, emission and ground state cross section emission of Tm^{3+} is shown in figure 8 [Aozasa et.al.2008,2002].

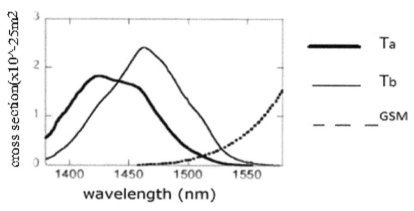

Fig. 8. Cross Sections of Tm^{3+}

The gain spectrum of TDFA is shown in figure 9 [Aozasa et.al.2008, 2002]. A gain of 22dB (approximate) is obtained from 1460-1485nm wavelength range. After this wavelength range the gain reduces sharply.

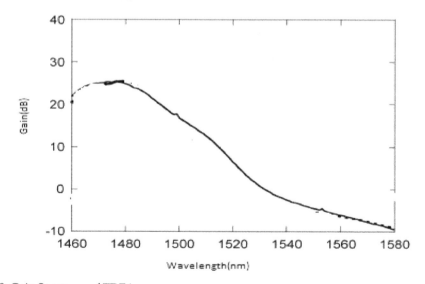

Fig. 9. Gain Spectrum of TDFA

So, to utilize the S-band, TDFA is proposed to be used by various authors. The S-band has attracted attentions because it has low fiber loss, low dispersion and also high gain and efficiency. The summary of the work done on EDFA, TDFA and TDFA-EDFA amplifiers is given in Table 3.

Type of Amplifier/ Parameters	EDFA in [Qiao & Vella2007]	TDFA [Aozasa et.al.2008]	TDFA-EDFA [Sakamoto et.al.2006]
Number of Stages	Two	Single	Four
ASE and its Correction Function	Considered with signal wavelength 1546nm	Considered	Not considered
Peak Gain /Range of Gain	0-37dB	22.6dB	20dB
Gain Excursion	0.35dB	0.35dB	2dB
Range of Input Power	-5 to 5 dBm	-32 to -2 dBm	-20 to -10dBm
Signal Gain Band	1525-1565nm	1479-1507nm	1460-1537nm
Noise Figure	Not considered	< 6.5dB	<7dB
No. of DWDM Channels Considered	8	8	4

Table 3. Summary of work done in [Qiao & Vella2007], [Aozasa et.al.2008], [Sakamoto et.al.2006]

8. Hybrid amplifiers

There is one more method of utilizing fiber amplifiers for optimum utilization of available fiber bandwidth i.e. by way of using various combinations of optical amplifiers in different wavelength ranges. The amplifiers can be connected either in parallel or in series. This configuration is termed as Hybrid Amplifier which is highly viable for the above discussed cause. In parallel configuration, the DWDM signals are first demultiplexed into several wavelength-band groups with a coupler, then they are amplified by amplifiers that have gains in the corresponding wavelength band and then they are multiplexed again with a coupler. The parallel configuration is very simple and applicable to all amplifiers. However, it has disadvantages also e.g. an unusable wavelength region exists between each gain band originated from the guard band of the coupler. Also, the noise figure degrades due to the loss of the coupler located in front of each amplifier. On the contrary, the amplifiers connected in series have relatively wide gain band, because they do not require couplers. Hybrid configurations can be made by combination of the following:

- **EDFAs and FRAs:** It has been observed that the gain spectrum of FRAs can be tailored by adjusting the pump powers and pump wavelengths. So this property is used to increase the amplification bandwidth of EDFA [Thyagarajan& Kakkar 2004; Oliveira et.al.2007, Kaur &Gupta 2008].

- **TDFAs and FRAs:** Combining FRAs with TDFAs is very effective approach, because FRAs can provide any gain bandwidth by selecting the appropriate pump wavelengths. However, a drawback with FRAs is that double Rayleigh scattering (DRS) degrades the amplified signals [Percival & Williams 1994; Komukai et.al.1995,2001; Royet.al.2002and Aozasa et.al.2002].
- **TDFAs and EDFAs:** Hybrid amplifiers consisting of all rare –earth –doped- fiber amplifiers are easier to utilize than those incorporating FRAs, because these are free from DRS. These hybrid amplifiers are relatively simple in gain spectra control [Sakamoto et.al.2006 and Kaur & Gupta 2010]. Hybrid doped fiber amplifiers with different gain bandwidths have attracted a large interest for increasing the transmission capacity of long haul wavelength multiplexed optical communication systems in C-band and L-band.

Out of these, one of the-state-of-art hybrid amplifiers is TDFA -EDFA configuration. It is observed that for TDFA-EDFA configuration, the total gain of hybrid amplifier is given as product of gain of TDFA and gain of EDFA. The gain bandwidth is extended by cascading EDFA with TDFA. When EDFA is cascaded with TDFA in series, the total gain of hybrid amplifier is given by product of individual gains of each amplifier.The gain of TDFA is given as

$$G_{T(\lambda)} = \exp\left[\left(\sigma_{T(1480)}NT_{2-\sigma T(1390)}NT_{1-\sigma T(800)}NT_0\right)\right]\left(\eta_T\,L_T\right) \tag{8}$$

The gain of EDFA is given as:

$$G_{E(\lambda)} = \exp\left[\left(\sigma_{E(1530)}NE_{2-\sigma E(1480)}NE_1\right)\left(\eta_E\,L_E\right)\right. \tag{9}$$

This means the total gain of hybrid amplifier is given as:

$$G_{(\lambda)}=G_{T(\lambda)}X\,G_{E(\lambda)} =$$
$$=\exp\left[\left(\sigma_{T(1480)}NT_{2-\sigma T(1390)}NT_{1-\sigma T(800)}NT_0\right)\left(\eta_T\,L_T\right)\right]X\exp\left[\left(\sigma_{E(1530)}NE_{2-\sigma E(1480)}NE_1\left(\eta_E\,L_E\right)\right]\right. \tag{10}$$

In the above equations, $\sigma_{T(1480)}$, $\sigma_{T(1390)}$ and $\sigma_{E(980)}$ denotes cross-sections of excited state absorption, stimulated emission and ground state emission of TDFA. Similarly, $\sigma_{E(1530)}$, $\sigma_{E(1480)}$ represents the respective cross sections of EDFA. η_T and η_E represents the confinement factors of TDFA and EDFA respectively.

The above stated mathematical equations clearly illustrate the fact that the gain of hybrid amplifier broadens from 1460 nm to 1530 nm wavelength range. Further there is a noticeable reduction in the noise figure correspondingly in the hybrid amplifier. This affects in the gradual increase in the number of transmission channels of DWDM system, thereby increasing the overall transmission capacity of the optical communication system. The statistical analysis of TDFA-EDFA hybrid amplifier and EDFA-TDFA hybrid amplifier is done. The configuration TDFA-EDFA means the hybrid amplifier in which TDFA is in first stage and EDFA is in second stage, whereas configuration EDFA-TDFA means EDFA is in first stage and TDFA is in second stage. For this analysis, it is assumed that both fibers have step refractive index homogeneously broadened spectrum of thulium and erbium ions. We consider three levels of

TDFA i.e. $_3H^6$, $_3H^4$ and $_3F^4$, the other two levels of TDFA i.e. $_3H^5$ and $_3F^{2,3}$ are ignored as the rate of their non radiation (τ_{nr}) to the corresponding levels are very high. Similarly, in case of EDFA, two levels $_4I^{15/2}$, $_4I^{13/2}$ is considered and level $_4I^{11/2}$ is ignored for the same reasons. From the absorption and emission spectra, it is clear that the absorption and emission peaks of EDFA coincides at 1530nm, while the absorption peak of TDFA lies at 1430nm and emission peak of TDFA lies at 1460nm. Form the gain spectrum of EDFA, from which it is clear that EDFA has peak gain at 1530nm, beyond which the gain reduces slightly and remains flat almost until 1550nm. After that gain reduces sharply. This gain can be flattened by cascading TDFA with EDFA. Fig. 10 shows the flattened gain spectrum of Hybrid amplifier by cascading TDFA and EDFA. The thulium doped fiber (TDF) in first stage was forward pumped with a 1390nm pump laser diode (LD) and erbium doped fiber (EDF) in second stage is pumped with a 980nm LD. For efficient amplification the concentration of TDF $^{+3}$ ions was kept very high (approx. 7500 ppm) [Percival & Williams 1994; Komukai et.al.1995]. Table 4 shows the different characteristics of both configurations.

Feature	TDFA-EDFA	EDFA-TDFA
Gain	25 dB for 1456nm-1556nm range	20 dB for 1485nm-1550nm range
Noise Figure	<6Db	<7dB

Table 4. Features of TDFA-EDFA & EDFA-TDFA Amplifiers

F-ratio is calculated for the parameters mentioned in above Table 4. The calculated and tabulated value of F-ratio is shown in Table 5.

Source of variation	D.F	SS	MS	F-ratio	Tabulated F-ratio (1,2)
Between samples	1	1.5	1.5	.05	18.51
Within Samples	2	39.6	29.9		
Total	3	41.1			

Table 5. F-Measure Results

The table 5 shows that the F-ratio is significant of 5% level which means that both hybrid configurations work differently. F-ratio is used to judge whether the difference among several sample means is significant or just a matter of sampling fluctuations. MS residual is always due to fluctuations of sampling and so serves as the basis for the significance test. The F-ratio is compared with its corresponding table value for the given degree of freedom at a specified level of significance. The table 5 shows that both the F-ratios are significant of 5% level which means that TDFA-EDFA amplifier work differently as compared with EDFA-TDFA amplifier. The gain spectrum of TDFA-EDFA is more widened as compared to that of EDFA-TDFA configuration. Since there is a large difference between the calculated and the table value of F. So, the null hypothesis is rejected. For WDM systems, TDFA-EDFA has a great impact as a hybrid amplifier as compared with EDFA-TDFA amplifier[Kaur & Gupta 2009]. As studied from the existing schemes, the amplification of DWDM signals using TDFA-EDFA hybrid amplifiers have major problems and short comings which are as listed below:

- The use of increasing number of channels in the present day DWDM optical networks requires a flat gain spectrum across the whole usable bandwidth. The unflattened gain spectrum of hybrid amplifiers implies that different channels of a DWDM system are amplified by different amounts. Hence a need is felt to broaden as well as flatten the gain spectrum of hybrid amplifier.
- It is observed that amplified spontaneous emission and its correction function for hybrid amplifiers have not been carried out leading to lesser gain and more noise of the signal. So, there is a need to analyze different parameters e.g. gain, noise figure, amplified spontaneous emission of hybrid amplifier and its correction function.
- No scheme or algorithm has been designed to allow the hybrid amplifier to maintain a constant output signal power, independent of the optical wavelength and input power level. There are many occasions when constant optical signal power, independent of input power, is more desirable e.g. in an optical preamplifier in an optical receiver and automatic power control cannot guarantee constant signal output power.
- Hybrid amplifiers proposed till dates are using four or more than four amplifiers to achieve desirable gain, leading to higher complexity, noise. Hence there is a dire need to minimize number of amplifiers in hybrid configuration.
- DWDM system till date are upto thirty two (32) channels but with lesser gain and high noise figure. The systems having adequate gain have been designed only upto eight (8) channels. Hence, there is a drastic need to increase the number of channels in DWDM system.

Although there are many improvements in gain spectrum of EDFA, but still the improved configurations are unable to provide enough bandwidth for emerging high quality parameters like gain, noise figure, amplified spontaneous emission etc. Therefore, there is a need to search for a new and versatile approach that enables an effective system with adequate bandwidth to accommodate large number of DWDM channels. One approach that can do the job is use of hybrid amplifier consisting of TDFA and EDFA in cascaded series combination. This hybrid amplifier is proven effective in DWDM systems. Several challenging points of research are realization and development of hybrid amplifiers, which can increase the bandwidth for S-band, C-band and L-band. The biggest challenge with hybrid amplifier is to maintain and offer high bandwidth in case of higher number of channels.

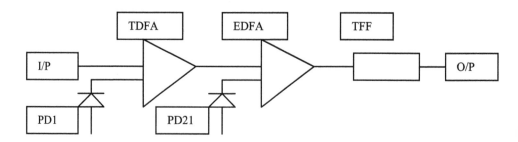

Fig. 10. Schematic Diagram of Cascaded TDFA-EDFA Hybrid amplifier using TFF

It is clear with the configuration as shown in figure 10 that a wide bandwidth spectrum of nearly 100nm i.e. from 1460nm to 1560nm wavelength range is obtained. This also includes the 1510nm-1520nm range where EDFA as well as TDFA has no large gain for themselves. This is also observed that this gain is unflattened mainly from 1520nm to 1540nm region. This whole wavelength range is flattened by using a seven layer interference filter (TFF).A seven layer optical thin film filters consists of a stack of seven dielectric thin a film is used along with a cascaded TDFA and EDFA [Kaur & Gupta 2010]. There are so many ways to flatten the gain bandwidth of EFDA such as gain equalizers based on Mach-Zehnder optical filters, interference filters or long period grating and fluoride or tellurite based EDFA. The figure (11) shows a schematic diagram of a seven layer dielectric interference filter for gain flattening of hybrid amplifier consisting of cascaded TDFA with EDFA. A seven layer dielectric film is proposed as a gain equalizer. In figure (11) the shaded layer as high index layer having refractive index 2.4. The unshaded layer is a low index layer having refractive index as 1.46. The refractive index of fiber is assumed as 1.46. The third and sixth layer of this filter is half wavelength thick and all other layers have one fourth wavelength thickness. The filter is so designed that transmission loss occurs around the maximum gain of hybrid amplifier i.e. at 1531nm. The transmission loss is about 9dB. The flattened gain bandwidth of hybrid amplifier with the help seven layer deictic filter is shown in figure (12). The gain variation is less than $\pm$ 2.5% in the wavelength region of 1460-1560 nm. A [2X2] square matrix of a dielectric filter for TM mode is given as

$$[Matrix]= [M] = \prod_{x=1} b \begin{bmatrix} cos\beta x & \frac{j}{q}sin\beta x \\ jqsin\beta x & cos\beta x \end{bmatrix} \tag{11}$$

The transmission of a TFF is given as

$$T= \left[1 + \frac{4R}{(1-R)2} sin^2\left(\frac{\Phi}{2}\right)\right]^{-1} \tag{12}$$

From equations (11) & (12), we get following equation for transmission

$$T= \left[\frac{2\,n\,fiber}{\left(m_{11}+m_{12}n_{fiber}\right)n_{fiber} + \left(m_{21}+m_{22}n_{fiber}\right)}\right]^2 \tag{13}$$

Where m_{ij} are the components of the matrix [M]. Here we assume n_{fiber}=1.46.Since the gain peak of hybrid amplifier occurs at 1530nm. So here we designed TFF such that the maximum transmission loss occurs at around the gain peak at wavelength 1530nm, which is observed as 9dB. In case of TE mode all parameters remain same except q_x is replaced by p_x. It is clear from formula given in equation (13) that designing of interference filters with desired wavelength spectrum and transmittance is possible by selecting proper of layer of dielectric films, their thickness and refractive indices of core and cladding of the optical fiber.

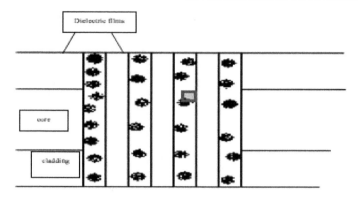

Fig. 11. Schematic Diagram of the dielectric multi-layer Interference Filter (TFF)

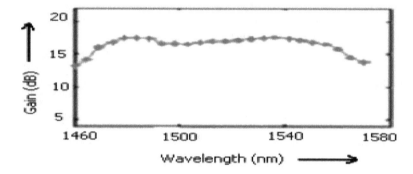

Fig. 12. Broadened and Flattened Gain Spectrum of Hybrid Amplifier

9. Conclusion

Wavelength multiplexing (WDM) technology along with optical amplifiers is used for optical communication systems in S-band, C-band and L-band. To improve the overall system performance Hybrid amplifiers consisting of cascaded TDFA and EDFA with different gain bandwidths are preferred for long haul wavelength multiplexed optical communication systems. It has found that calculated value of F ratio is very much different from the tabulated value, so the difference between parameters is considered as significant and we reject the null hypothesis. Here, we are able to conclude that for WDM systems, TDFA-EDFA hybrid fiber doped amplifier has higher gain and lower noise figure. So, this configuration gives better performance in WDM systems as compared with the EDFA-TDFA hybrid configuration. With this design it has also been found that when TDFA with EDFA are cascaded in series then gain spectrum is broadened. The gain variation is less than ± 2.5% in the wavelength region of 1460-1560 nm. The TF filter is so designed that transmission loss occurs around the maximum gain of hybrid amplifier i.e. at 1531nm. The transmission loss is about 9dB.The simulation process can be represented by a flowchart as shown in figure 13.

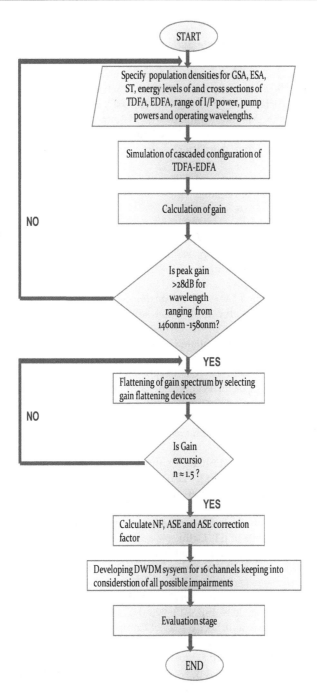

Fig. 13. Flowchart Showing Simulation Process

10. References

Agarwal Govid P., *"Fiber Optic Communication Systems"*, John Wiley & sons, Inc. Publication, 2003.

Aozasa S., H.Masuda, T. Sakamoto, K.Shikano and M.Shimizu, "Gain-Shifted TDFA Employing High Concentration Doping Technique with High Internal Power Conversion Efficiency of 70%,"*Electron. Letters*, Vol. 38, No. 8, pp. 361-363,2002.

Aozasa S, Hiroji Masuda, Makoto Shimizu and Makoto Yamada, "Novel Gain Spectrum Control Method Employing Gain Clamping and Pump Power Adjustment in Thulium- Doped Fiber Amplifier" *Journal of Lightwave Tech.* Vol. 26, No.10, May 2008.

Arbore .M.A., Y. Zhou, G.Keaton and T.Kane, "34 dB Gain at 1500nm in S- Band with Distributed ASE Suppression," presented at Eur. Conf. Opt. Commun., Denmark, Sept. 2002.

Bakshi B., M.Vaa, E.A.Golovchenko, H.Li and G.T. Harvey, " Impact of Gain Flattening Filter Ripple in long Haul WDM Systems", *Proc. 27th Eur. Conference on Optical Communication 2001* Vol. 3, pp. 448-449, Sept. 2001.

Becker P.C, Olsson N.A. and. Simpson J.R, *"Erbium Doped Fiber Amplifiers"*. New York: Academic, 1999, Chapter 7.

Desurvire Emmanuel, *"Erbium Doped Fiber Amplifiers- Principles and Application"*, Hoboken, NJ: John Wiley & Sons, Inc. ISBN 0-471-58977-2,2002 Chapter 5.

Hwang B. C. et al., "Cooperative Up- Conversion and Energy Transfer of New High Er^{+3} and Er^{+3} – Yb^{+3} Doped Phosphate Glasses," *Journal of Optical Society of America*, Vol. 17, No. 5, pp. 833-839, 2000.

Kasamatsu T., Yano Y. and Ono T., "Gain-Shifted Dual –Wavelength- Pumped Thulium-Doped Fiber Amplifier for WDM Signals in the 1.48-1.51-μm Wavelength Region," *IEEE Photonics Technology Letters*, Vol. 13, No. 1, pp.31-33,2001.

Kaur Inderpreet, Gupta Neena, "Statistical Analysis of Different Configurations of Hybrid Doped FiberAmplifiers", *International Journal of Electrical and Electronics Engineering* 3:8 2009 pp 515-520.

Kaur Inderpreet, Gupta Neena, "Enhancing the Performance of WDM Systems By Using TFF in Hybrid Amplifiers", 2010 *IEEE 2nd International Advance Computing Conference*, Patiala, India, pp 106- 109, Feb 2010.

Kaur Inderpreet, Gupta Neena,, "Increasing the Amplification Bandwidth of Erbium Doped Fiber Amplifiers by Using a Cascaded Raman-EDFA Configuration", *Photonics 2008*, P.284, Dec. 2008 at IIT, Delhi.

Kawai Shingo, Masuda Hiroji, Suzuki Ken Ichi, Aida Kazuo, "Wide Bandwidth and Long Distance WDM Transmission Using Highly Gain Flattened Hybrid Amplifier, *IEEE Photonics Technology Letters*, Vol. 11, pp. 886-888, July 1999.

Keiser Gerd, *"Optical Fiber Communications"*, 4th Edition, Tata McGraw-Hill Education Pvt. Ltd., New Delhi, Inc. 2009, ISBN-13: 978-0-07-064810-4.

Kemtchou J., M. l, Chatton F. and Lecoy T. G., "Comparision of Temperature Dependences of Absorption and Emission Cross – Section in Different Hosts of EDFAS," *Proc. Opt. Amplifiers Appl.*, 1996, pp. 126-129, Paper FD2.

Kim Hyo Sang, Yun Seok Hyun, Hyang Kyun, Kim Namkyoo Park and Kim Byoung Yoon, " Actively Gain-Flattened EDFA Over 35nm by Using All-Fiber Acousto-Optic Tunable Filters", *IEEE Photonics Technology Letters*, Vol. 10, pp. 790-792, June1998.

Komukai T., Yamamoto T., Sugawa T andMiyajima Y, "Upconversion Pumped Thulium-Doped Fluoride Fiber Amplifier and Laser Operating at 1.47µm," *Journal of Quantum Electronics*, Vol. 31, pp. 1880-1889, Nov. 1995.

Lee Y.W, Nilsson J, Hwang S.T. and Kim S.J., "Experimental Characterization of a Dynamically Gain –Flattened EDFA", *IEEE Photonics Technology Letters*, Vol. 8, No. 12, pp. 1612-1614, Dec.1996.

Lobo Audrey Elisa, Besley James A. and C.Martijin De Sterke, " Gain Flattening Filter Design Using Rotationally Symmetric Crossed Gratings", *Journal of Lightwave Tech.* Vol. 21, No. 9, pp. 2084-2088, 2003.

Lu Yi Bin and Chu P.L., "Gain Flattening by Using Dual-Core Fiber in Erbium Doped Fiber Amplifier",*IEEE Photonics Technology Letters*, Vol. 12, No. 12, pp. 1616-1617, Dec. 2000.

Mynbaev D.K, L.L.Schiner' *"Fiber Optics Communications Technology"*, Pearson Education, Delhi, Inc.,2003, ISBN 81-7808-317-5.

Oliveira J.C. .Silva R.F., Rossi S.M, Rosolem J.B. and .Bordonalli A.C, "An EDFA Hybrid Gain Control Technique for Extended Input Power and Dynamic Gain Ranges with Suppressed Transients", *IMOC 2007*, pp. 683-687, 2007 SBMO/IEEE MTT-S.

Olsson N.A., *"Lightwave Systems with Optical Amplifiers"*, Journal of Lightwave Tech. Vol. 7, July 1989.

Ono Hirotaka, Yamada Makoto and Ohishi Yasutake, "Broadband and Gain Flattened Amplifier Composed of a 1.55µm- Band and a 1.58 µm- Band Er^{3+}- Doped Fiber Amplifier in a Parallel Configuration", *Electronics Letters*, Vol. 33, No. 8, pp. 710-711, May 1997.

Park Seo Yeon, Kim Hyang Kyun, Lyu Gap Yeol, Sun Mo Kang and Sang Yung Shin, " Dynamic Gain and Output Power Control in a Gain-Flattened EDFA", *IEEE Photonics Technology Letters*, Vol. 10, No. 6, pp. 787-789, June1998.

Pasquale F. D. and Federighi M., "Modeling of Uniform and Pair –Induced Up Conversion Mechanisms in High-Concentration Erbium Doped Silica Waveguides," *Journal of Lightwave Technology*, Vol.13, pp. 1858-1864, 1995.

Percival R.M. and Williams J.R, "Highly Efficient 1.064µm Upconversion Pumped 1.47µm Thulium Doped Fluoride Fiber Amplifier," *Electron Letters*, Vol. 30, No. 20, pp. 1684-1685, June 1994.

Qiao Lijie and Vella Paul J., "ASE Analysis and Correction for EDFA Automatic Control," *Journal of Lightwave Tech.* Vol. 25, No.3, May 2007.

Roy F., Sauze A., Baniel P. and Bayart D., "0.8-µm +1.4µm Pumping for Gain Shifted TDFA With Power Conversion Efficiency Exceeding 50%," Presented at *the Opt. Amplifiers Appl.* Vancouver, BC, Canada, Jul.2002, PD4.

Sakamoto Tadashi, Aozasa Shin- ichi, Yamada Makoto and Shimizu Makoto, "Hybrid Fiber Amplifier Consisting Of Cascaded TDFA and EDFA for WDM Signals", *Journal of Lightwave Tech.* Vol. 24 No.6 , June 2006.

Sohn Ik-Bu, Baek Jang Gi, Lee Nam Kwon, Kwon Hyung Woo and Song Jae Won, "Gain Flattened and Improved EDFA Using Microbending Long-Period Fiber Gratings", *Electronics Letters*, Vol.38, No. 22, pp. 1324-1325, Oct. 2002.

Sun .Y, Judkins J.B., Srivastava A.K., Garrett L.,J.L.Zyskind, Sulhoff J.W, Wolf C., Derosier R.M., Gnauck A.H., R.W.Tkach, J.Zhou, Espindola R.P., Vengsarkar A.M and Chraplvy A.R., "EDFA Transmission Response to Channel Loss in WDM Transmission System", *IEEE Photonics Technology Letters*, Vol. 9,No. 3, pp. 386-388, March 1997.

Y. Sun, Zyskind J.L. and Srivastava A. K., "Average Inversion Level, Modeling and Physics of Erbium Doped Fiber Amplifiers," *Journal of IEEE Sel. Topics Quantum Electronics.*, Vol. 3, no. 4, pp. 991-10007, Aug.1997.

Thyagarajan K., Kakkar Charu, "Novel Fiber Design for Flat Gain Raman Amplification Using Single Pump and Dispersion Compensation in S-Band", *IEEE Photonics Technology Letters*, Vol.22, pp. 2279-2286, Oct. 2004.

Yun Seok Hyun, Lee Bong Wan Lee, Kim Hyang Kyun and Kim Byoung Yoon, "Dynamic Erbium Doped Fiber Amplifier Based on Active Gain Flattening with Fiber Acousto-Optic Tunable Filters", *IEEE Photonics Technology Letters*, Vol. 11, No. 10, pp. 1229-1231, Oct. 1999.

5

The Least Stand-By Power System Using a 1x7 All-Optical Switch

Takashi Hiraga and Ichiro Ueno
National Institute of Advanced Industrial Science & Technology
Japan

1. Introduction

According to the explosive popularization of the internet, the infrastructure for a broadband optical network becomes very important. The most serious problem, however, for a broadband network is large energy consumption of the infrastructure. For example, over 1% of generated energy is consumed in Japan, and in the near future it will become over 10 %. An energy-saving measure, therefore, is energetically promoted in all of Japan including the government, industry and universities. Namely, an effort to save energy in homes is important, because about 50 million families are living in Japan. With the spread of "fibre to the home (FTTH)", in the future, several services such as high definition television (HDTV), will use the Internet, which will become a larger energy consumer in homes. Equipment for FTTH, which include electrical-switched devices, such as HUBs and Routers, are on stand-by for communication. If the energy consumption in every home decreased by 10%, it will bring about surprising energy savings.

An optically gated optical switch without any electric parts is composed of several lenses and a dye-dissolved high-boiling-point solvent, where the absorbance of the dye for the signal light is lower than 0.1 and for the gating light is over 3. The most versatile advantage of this system is easy selection of the wavelengths for both the gating light and the signal light. The signal light, which is transparent to the dye-solution, is refracted by a temporally formed microscopic thermal-lens (region with lower refractive index) that is locally heated around a focal point by the irradiation of the gating light (Tanaka et al., 2007; Tanaka et al., 2010, Ueno et al., 2003, Ueno et al., 2007, Tanaka et al., 2010).

Here, we have developed a local telecommunications system for FTTH using an optically gated optical switch composed of only optical parts. This system is suitable for a local-area network within a home. An optical-fibre line from a telephone office is directly connected to a 1x7 optically gated optical switch (Fig. 1). A terminal unit (optical interface; Opt-I/F) of the present system connected directly between a 1x7 optically gated optical switch (all-optical switch) and a PC, TV, IP-phone and so on, plays the role of a controller of 1x7 optically gated optical switch for the light-path switching of a PC, and so on. This unit sends a command to another terminal unit for negotiation among the terminal units using a 980 nm line. All terminal units connected to a 1x7 optically gated optical switch via a reflection-type optical star coupler in a 980 nm line can establish completely independently collision-free communication among the terminal units.

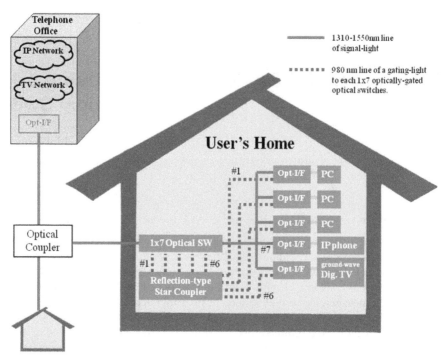

Fig. 1. Block diagram of the present system. The solid line is the optical fibre for data communication (wavelength of 1310-1550 nm), and the dashed line is the optical fibre for the control of a 1x7 all-optical switch (wavelength of 980 nm). The 1x7 optically gated optical switch (1x7 all-optical switch) is connected directly to the optical interface (Opt-I/F).

2. Current system

The current system concerning the least stand-by power by using a 1x7 all-optical switch is composed of a 1x7 optically-gated optical switch, 7 optical interfaces for each user terminal and a reflection-type star coupler (Fig. 1). The key equipment of the least stand-by power system is both a control system and an optically-gated optical switch, which is controlled directly by each user terminal ordering to occupy the circuit by selecting the 1x7 optically-gated optical switch. To avoid collisions among the user terminals, a control system is employed.

2.1 1x7 all-optical switch

The most advanced feature of an optically gated optical switch without any electrical parts is an easy selection of both the gating light and the signal light by selecting a suitable dye for a transparent wavelength region as the signal light and a large absorption wavelength region as the gating light. This switch is composed of four important optical components. The 1st component involves two-types of 7-bundled optical-fibres, both for incidence and collecting. The incidence optical fibre is composed of a central optical fibre for the signal light and six outer optical fibres for the gating light. All of the collecting optical fibre is for the signal light (details of these fibres are explained in section 2.1.3). The 2nd is a dye cell made of quartz filled with a high-boiling-point solvent and a dissolved dye for the absorbing gating light. The medium for operating an optical switch is perfectly dehumidified and deoxygenated, which brings long-life operation, even under light-irradiation (details of preparation are explained in section 2.1.2). The 3rd is a prism of a hexagonal truncated pyramid, located between a collecting lens for the dye-cell and a collimating lens for the collecting fibre. This prism brings a higher coupling efficiency to the collecting fibre. The 4th are lenses focusing both the signal light and the gating light from an incidence optical fibre to a dye-cell. Another two pairs of lenses collect signal light from the dye-cell, and focuses it to the collecting optical fibre.

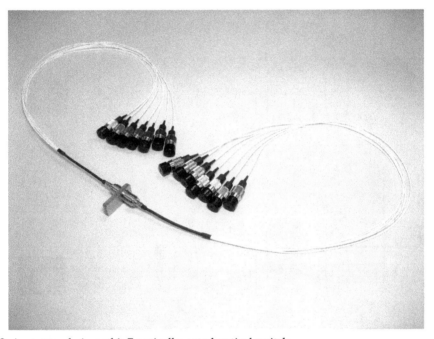

Fig. 2. An external view of 1x7 optically-gated optical switch.

The typical performance of the present 1x7 optically gated optical switch is summerized in Table 1. Both the insertion loss and the crosstalk are the mean value of the 7 exit ports.

Parameters	Specifications
Wavelength	1310-1490 nm
Gating light	980 nm
Insertion loss	6.5 dB min.
Crosstalk	34 dB min.
Dimension	95×30×12mm

Table 1. Typical performance of a 1x7 optically gated optical switch using different axis configurations under a gating light power of 20~35 mW.

2.1.1 Principle of an optically gated optical switch

We have developed two types of optically gated optical switches: coaxial configuration of the signal and gating light (Tanaka et al., 2007, Ueno et al., 2007), and a different axis configuration of the signal and gating light (Ueno et al., 2010). The former having a coaxial configuration can operate under a lower gating power, because the signal light is refracted at a region just heated by the gating light. This configuration possesses, however, a lower coupling efficiency to a collecting optical fibre because of ring-shaped refracted light. The later configuration of the different axis type is the currently adopted where the 1x7 optically gated optical switch (shown in Fig. 2) possesses large advantages. The basic operation of a different axis type switch is explained using a 1x2 optical switch (Fig. 3).

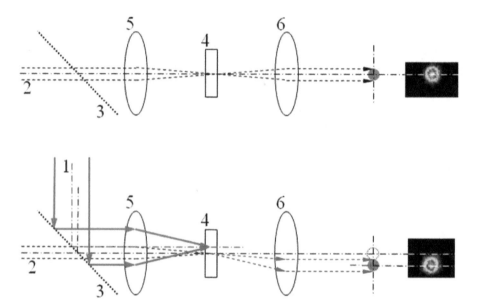

Fig. 3. Basic operation of a different axis configuration. Gating light (1) is mixed with the signal light (2) using a dichromatic mirror (3), and is introduced into a dye cell (thermal-lens forming element)(4) via a focusing non-spherical lens (5). The gating light and the signal light have different axis configurations. The switched light is collimated by a lens (6). In the case with gating light, the switched light is projected to a shifted position.

The reason why both the signal light and the gating light simultaneously pass through a dye-cell is due to the dichromatic property of the dye solution. An organic dye, generally, exhibits a relatively sharp absorption spectrum. An organic dye of "YKR3080", for example, exhibits both large absorption around a wavelength of 1000 nm and small absorption longer than 1200 nm, as shown in Fig. 4. This dichromatic property of an organic dye allows easy selection of both the signal and gating wavelength of light for present in the optically gated optical switch.

Absorbance
4
3.5
3
2.5
2
1.5
1
0.5
0
800
1000
1200
1400
1600
1800
Wavelength [nm]
Gating light [980 nm]
Signal light [1550 nm]

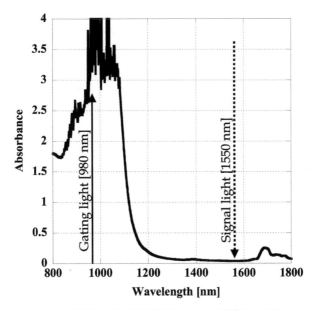

Fig. 4. Absorption spectrum of 0.2 wt%-YKR3080/solvent "S" in a 0.5 mm quartz cell for a 980 nm gating. The wavelength of the signal light is 1310-1550 nm.

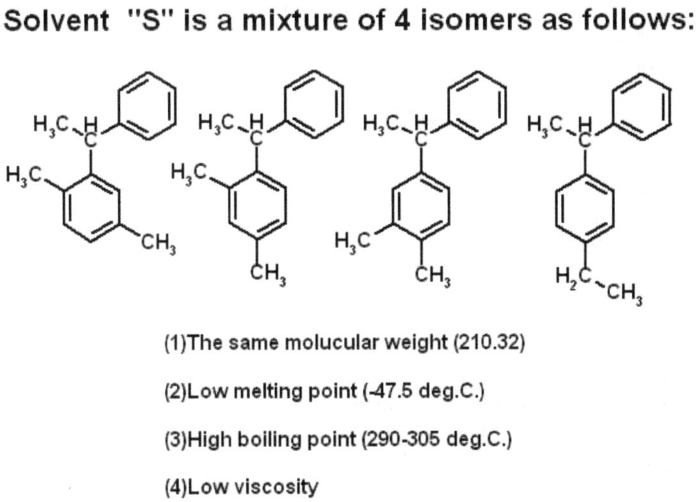

Solvent "S" is a mixture of 4 isomers as follows:
(1)The same molucular weight (210.32)
(2)Low melting point (-47.5 deg.C.)
(3)High boiling point (290-305 deg.C.)
(4)Low viscosity

Fig. 5. Solvent "S" is composed of 4 kinds of structural isomers with the same molecular weight: 1-Phenyl-1-(2,5-xylyl)ethane, 1-Phenyl-1-(2,4-xylyl)ethane, 1-Phenyl-1-(3,4-xylyl)ethane and 1-Phenyl-1-(4-ethylphenyl)ethane.

Another important component of the dye solution is the solvent. Here, solvent "S" is employed for a medium to form a thermal lens in the dye cell. When the gating light is focused into a dye solution, molecules of YKR3080 absorb energy of the irradiated light, which causes an increase in the temperature of the dye by thermal relaxation of the excited state. The transferred energy from the dye-molecule to solvent "S" (Fig. 5) around the focal point of the gating light forms a high-temperature region around the focal point of the gating light. As a relation between the refractive index (n) of solvent "S" and the temperature (T; deg.) is expressed as Equation (1); the refractive index of solvent "S" at the heated region become lower than the surrounding region (Hiraga et al., 2008).

$$n = 1.5742 - 0.00048259\ T \qquad (1)$$

This is the reason the signal light is refracted at around the focal point of the gating light. It is difficult to determine the exact shape of the region at a lower refractive index, because the dye solution has large absorbance at around 3 and the "Beer-Lambert law". We suppose the shape to be a triangular pyramid as shown in Fig. 6. We are now trying to measure the local temperature using the Raman-scattering method (Hiraga et al., 2010). In the case of different-axis configuration, the axis of the gating light (1) to be at about 40 µm parallel translation from the signal light (2) as shown in Fig. 6.

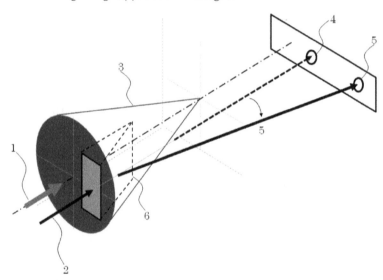

Fig. 6. Refraction of the signal light by an optical gating under different axis configurations of the gating light with the signal light: 1, gating light; 2, signal light; 3, thermal lens; 4, non-refracted light without gating light; 5, refracted light with gating light; 6, imaginary wedge-shaped thermal lens.

Fig. 7 shows a schematic drawing of the optics of the 1x7 optically gated optical switch. It is composed of 6 kinds of optical parts. The incidence 7-bundled optical fibre (1) is employed for both the signal light and the gating light, where the centre-to-centre distance of the core is 40 µm. The collecting 7-bundled optical fibre (6) is employed for only signal light, where centre-to-centre distance of the core is 250 µm. A prism of a hexagonal truncated pyramid (4) set up between the focusing lens (5) and the collecting lens (6) is employed for enhancing the coupling efficiency to a collecting 7-bundled optical fibre (6).

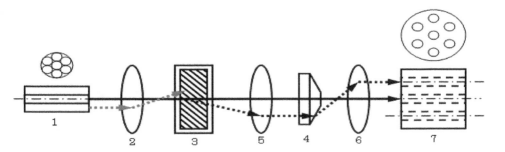

Fig. 7. Schematic drawing of the optics of the 1x7 optically gated optical switch: 1, incidence 7-bundled optical fibre; 2, focusing lens; 3, dye solution; 4, prism of hexagonal truncated pyramid; 5, focusing lens; 6, collecting 7-bundled optical fibre.

2.1.2 A dye-cell as one of the most important parts

The most important residual component for degradation of the dye solution in a quartz cell is both oxygen and water. Therefore, a dye-cell made of a quartz capillary is prepared in a vacuum glove box. First, a solvent "S" is dried using a molecular sieve for several hours. Second, a freeze-pump-thaw treatment of a dye-dissolved solution is employed several tens of times in a vacuum glove box. Next, a quartz capillary of outer diameter of 1 mm, inner diameter of 0.5 mm and 25 mm length is sealed at one end after being dried in a vacuum groove box. A dye of YKR3080 dissolved in a solvent "S" with a high concentration near saturation is charged into the capillary by a micro-syringe. Another end of the capillary is sealed temporally by glue in the vacuum glove box. At the outside of the glove box, another end of the capillary is sealed by melting quartz using a micro burner (Fig. 8).

Fig. 8. Micro cell made of a quartz capillary: outer diameter of 1 mm, inner diameter of 0.5 mm and 25 mm length. Both the purification and sealing processes are performed in an ultra-pure vacuum glove box.

2.1.3 A 7-bundled optical fiber

Two types of a 7-bundled optical fibre are developed: one is for incidence and the other is for collecting. The incidence optical fibre is composed of a centre fibre for the signal light and 6-outer fibres for the irradiating gating-light to a dye cell. The diameter of the clad of a single mode optical fibre for optical communication is reduced from 125 μm to 40 μm, so as to reduce the centre-to-centre distance of the core to 40 μm in a 7-bundled optical fibre as shown in Fig. 9(a); 7 optical fibres are assembled as the closest packing configuration in a sleeve of inner diameter 120 µm.

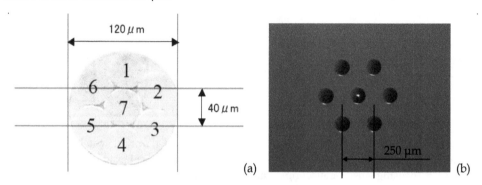

Fig. 9. Incidence (a) and collecting (b) 7-bundled optical fibre composed of a single-mode optical fibre (core diameter of about 10µm). Centre fibre of (b) is lightened.

The collecting optical fibre is composed of 7 optical fibres for the collecting signal light, where the centre-to-centre distance of the core is 250 µm; 7 optical fibres with a clad diameter of 125 µm are independently set up in a ferrule with 7 holes, as shown in Fig. 9(b).

2.1.4 A prism of hexagonal truncated pyramid for improving the throughput to the receiving optical fiber

A 3 mm diametre prism of the hexagonal truncated pyramid set up between the focusing lenses is employed for enhancing the coupling efficiency to a collecting 7-bundled optical fibre (Fig. 10). This prism brings rectangular incidence of refracted light from a dye-cell to a collecting optical fibre. The non-refracted signal light, i.e., in the case without the gating light, passes through a flat area of the centre top in the prism. The refracted signal light, in the case with gating light, passes through the side slope in the prism (one of the 6-polished surfaces).

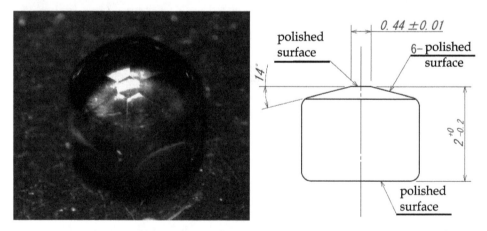

Fig. 10. Prism of a hexagonal truncated pyramid made by a glass-mould pressing method.

2.2 The least stand-by power system

The optical-fibre line from a telephone office is directly connected to a 1x7 optically gated optical switch (Fig. 1). A terminal unit (optical interface; Opt-I/F) of the present system is connected directly between a 1x7 optically-gated optical switch and a PC, TV, IP-phone and so on. It serves as a controller of the 1x7 optically gated optical switch for light-path switching of the PC and so on. This unit sends a command to another terminal unit for negotiation among the terminal units using a line having a wavelength of 980 nm. All terminal units connected to a 1x7 optically gated optical switch via a reflection-type optical star coupler in the 980 nm line can establish completely independent collision-free communication among these terminal units.

Several functions of optical LAN of the present system connected to a telephone office are as follows (Fig. 11). The Ethernet of the electrical and optical performance is Tx of 1550 nm and Rx of 1310 nm, Filtering of data communication involves 1000 base-T, 100 base-T and serial communication, Communication of the control system for data communication, and The Measurement of energy consumption for each port.

Several functions of optical LAN of the present system connected to a PC, TV, IP-phone and so on are as follows. The Ethernet of the electrical and optical performance is Tx of 1310 nm and Rx of 1550 nm, Filtering of data communication involves 1000 base-T, 100 base-T and serial communication, Communication of the control system for data communication, The Measurement of energy consumption for each port, and Gating-light of 980 nm for controlling a 1x7 all-optical switch. Function of a reflection-type optical star coupler in 980 nm line is delivery of gating-light to each 1x7 optically-gated optical switches.

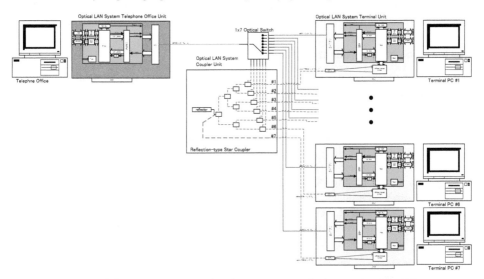

Fig. 11. Overview of the present system for demonstration. Left upper is a unit for establishing an IP Network、a TV network, and so on in a home network. Top middle is "Reflection-type Star Coupler" with the 1x7 optically-gated optical switch. Right blocks are 7 downstream units from the 1x7 optical switch located in a user's home.

An "Optical LAN System as a Telephone Office Unit" and a "Telephone Office PC" (left upper in Fig. 11) are a virtual unit for establishing an IP Network 、 TV network, and so on in the home network; 7 downstream units from a 1x7 optical switch are settled in a user's home. An "Optical LAN System Coupler Unit" (top middle in Fig. 11) consists of a "Reflection-type Star Coupler", which carries both communication among each "Terminal PC" and control of the 1x7 optical switch distributing gating light of 980 nm in a user's home. Each "Terminal PC" communicates with each other using gating light of 980 nm via the "Optical LAN System Terminal Unit". When one "Terminal PC" is ready to communicate using the circuit (a connected circuit is free from exclusiveness), an order is sent to select the 1x7 optically-gated optical switch using 980 nm light. Then, the circuit between the "Optical LAN System as a Telephone Office Unit" and the "Terminal PC" are occupied and begin to communicate. Using the present system, the idling "Terminal PC" is to shut down, which allows the least stand-by power system.

2.2.1 Setup of the system

The functional block of the present system in a user's home, and categorised into three groups as shown in Fig. 12. The 1st is a continuously power supplied unit, such as a telephone, where a relatively slow transmission speed is allowed (blue block in Fig. 12). The 2nd is medium transmission-speed terminals, indicated as yellow, where both e-mail and network surveying is mainly used. The 3rd is high transmission-speed terminals, indicated as pink, where a dynamic (or moving) image is mainly used. The "Optical LAN System Telephone Office Unit" with a low speed I/F, which is always powered by the least energy takes part in controlling these three units. The other unit of both "Middle Speed I/F" and "High Speed I/F" is always stand-by, and working when communication is needed.

2.2.2 Operation of the present system

The initial state of the present system is through connection from the "Optical LAN System Telephone Office Unit" (left upper) to the "Optical LAN System Terminal Unit" (low speed I/F : blue block at right lower; continuously power supplied). Without any gating light in a 1x7 optically-gated optical switch, the optical path is automatically selected from an incidence port to a central port of the collecting optical fibre (exit side). The central port of the exit side in a 1x7 optically gated optical switch is connected to an "Optical LAN System Terminal Unit" (blue block at right lower; continuously power supplied). At this time, it is able to occupy a circuit using 980 nm light delivery from an "Optical LAN System Terminal Unit" to the other "Optical LAN System Terminal Unit" via an "Optical LAN System Coupler Unit". By means of these negotiations, an order for occupying the circuit from each "Optical LAN System Terminal Unit" will be arranged by the "Optical LAN System Terminal Unit" (low speed I/F: blue block at right lower; continuously power supplied).

An "Optical LAN System Terminal Unit#1"(right upper one) output gating light from a 980 nm LD in the "Optical LAN System Terminal Unit#1" is used to occupy a circuit for communication. An "Optical LAN System Coupler Unit" delivers gating light of 980 nm from the "Optical LAN System Terminal Unit#1" to both a "1x7 Optical Switch" and an "Optical LAN System Terminal Unit" (low speed unit), which establish communiation of the "Optical LAN System Terminal Unit". After finishing communication (data transfer) against for "Optical LAN System Telephone Office Unit", the "Optical LAN System Terminal Unit#1" broadcast to the "Optical LAN System Terminal Unit" end of communication using 980 nm light, releases the occupied circuit. By these procedures, the connection of the circuit becomes the initial state shown in Fig. 12. By the alternative action described above, each "Optical LAN System Terminal Unit"(#1~#6) can communicate alternatively to the "Optical LAN System Telephone Office Unit".

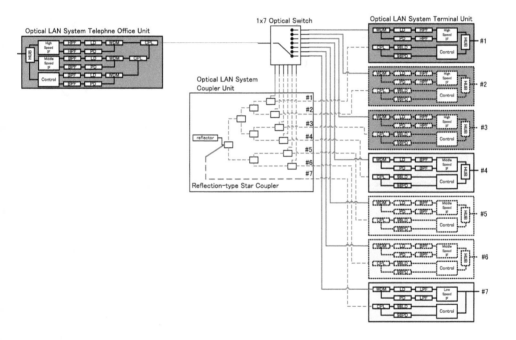

Fig. 12. Full system of the innovative least stand-by power equipment. The blue block at the right lower indicates a unit continuously supplied with power (low speed I/F). Three pink blocks at the right upper (high-speed I/F) and three yellow blocks at the right middle (middle-speed I/F) indicate users. The white block at the centre indicates a reflection-type star-coupler. The white block at the upper centre indicates an optically-gated optical switch. The pink block at the left upper indicates a unit in a telephone office.

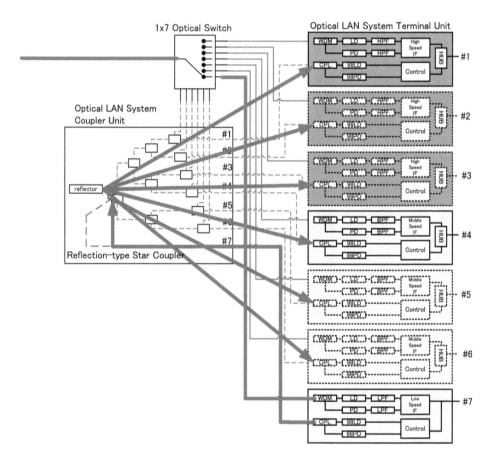

Fig. 13. First step of communication among an "Optical LAN System Terminal Unit" using 980 nm light.

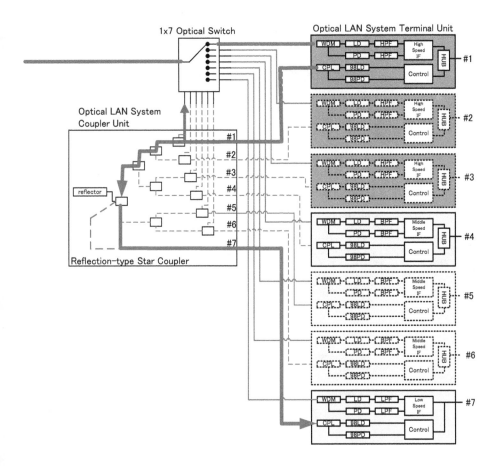

Fig. 14. Operation of communication between one "Optical LAN System Terminal Unit" and an "Optical LAN System Telephone Office Unit" using 980 nm light.

In Fig. 15, an external view of "The least stand-by power system using a 1x7 optically-gated optical switch" currently being developed is shown. The specifications of both the OLT (Optical Line Terminal) and the ONU (Optical Network Unit) are summarized in Table 2 and Table 3, respectively.

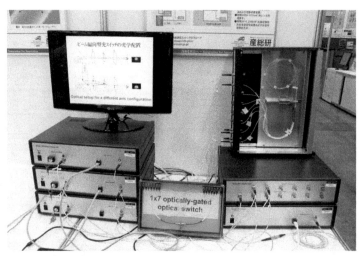

Fig. 15. External view of "The least stand-by power system using a 1x7 all-optical switch".

E/O Unit		
Electrical Interface	OLT Tx+/-	1250Mbps DATA_IN LVPECL SMA NRZ
Optical Interface	Wavelength	1310-1490 nm
	Output Power	≧+3.5dBm@Enable≦-39dBm@Disable
	Extinction Ratio	≧9dB
O/E Unit		
Electrical Interface	OLT RX+/-	1250Mbps DATA_OUT LVPECL SMA NRZ
Optical Interface	Wavelength	1310-1490 nm
	Min. Input Power	≦-30.0dBm
	Max. Input Power	≧-8.0dBm
	LOS Detection	≧-45.0dBm(Detection)≦-30.0dBm(Release)
Function		
Display	LED	Power Supply, LD, LOS
Switch	SW	Power Supply, LD
Lock	Key	LD Source
Others		
Controll Interface	USB	Rev. 2.0 full speed USB-B
	Controll Item	LD ON/OFF LOS Monitor
Power Supply	-	DC 5V, 0.8 W
Size	-	W210mm × D350mm × H99mm
Weight	-	2.5 kg

Table 2. Specifications of OLT (Optical Line Terminal) Optical Interface Unit.

E/O Unit		
Electrical Interface	ONU Tx+/-	1250Mbps DATA_IN LVPECL SMA NRZ
	Burst Enable	LVTTL High-Z Active-High
Optical Interface	Wavelength	1310-1490 nm
	Output Power	$\geqq$-0.5dBm@Enable$\leqq$-45dBm@Disable
	Extinction Ratio	$\geqq$9dB
O/E Unit		
Electrical Interface	ONU RX+/-	1250Mbps DATA_OUT LVPECL SMA NRZ
Optical Interface	Wavelength	1310-1490 nm
	Min. Input Power	$\leqq$-26.0dBm
	Max. Input Power	$\geqq$-3.0dBm
	LOS Detection	$\geqq$-44.0dBm(Detection)$\leqq$-26.0dBm(Release)
Function		
Display	LED	Power Supply, LD, LOS
Switch	SW	Power Supply, LD
Lock	Key	LD Source
Others		
Controll Interface	USB	Rev. 2.0 full speed USB-B
	Controll Item	LD ON/OFF LOS Monitor
Power Supply	-	DC 5V, 0.8 W
Size	-	W210mm × D350mm × H99mm
Weight	-	2.5 kg

Table 3. Specifications of ONU (Optical Network Unit) Optical Interface Unit.

3. Conclusion

A versatile optically gated optical switch using an organic dye is suitable for light-path switching in optical communication at various wavelengths of the signal light. The feature of utilizing a thermal-lens effect generated in a thin layer of a dye solution is possible to cover a wide range of wavelengths by varying the types of pigment. An organic dye for absorbing the gating light combined with a high-boiling solvent for forming a thermal lens is properly selected by choosing the wavelengths of both the gating light and the signal light. The formed thermal-lens in a dye solution is very small, which brings about high-speed switch of around 1 msec.

An epoch-making 1x7 optically-gated optical switch has been developed using both a 7-bundled optical fibre with a centre-core distance of 40 µm and a prism of a hexagonal truncated pyramid. Around 0.6 of the coupling efficiency of incidence light to a 7-bundled optical fibre and smaller than 40 dB of crosstalk for both incidence light have been achieved. Using the 1x7 optically-gated optical switch, we have developed a least stand-by power system as well. This system is located in a home, and is composed of the 1x7 optically gated optical switch and a reflection-type optical star coupler, which directly connects between an optical coupler from a telephone office and a PC, TV, IP-phone, etc. Using this system, the least stand-by power in the home will be achieved.

4. Acknowledgment

This work is a joint study of Norio Tanaka and Shigeru Takarada of Dainichiseika Color & Chemicals Mfg. Co., Ltd., Noriyasu Siga and Kazuya Ohta of Trimatiz, Ltd., Hirofumi Watanabe, Toshimasa Tamura and Keiji Negi of Inter Energy Co., Ltd. and Masaki Kubo of HIMEJI RIKA CO., LTD.. This study is supported by NEDO.

5. References

Hiraga, T.; Ueno, I.; Nagaeda, H.; Shiga, N.; Watanabe, H.; Futaki, S. & Tanaka, N. (2008). Development of an Optically gated Optical Switch using an Organic Dye - Applied to Local Telecommunication Technology. *Proceedings of SPIE*, Vol. 6891, No. 68910G, PP. 1-15.

Hiraga, T.; Ueno, I.;Siga, N.; Watanabe, H.;Futaki, S.; Kubo, M.; Takarada, S.;Tanaka, N. (2011). Switching mechanism of an optically-gated optical switch using an organic dye. *Proc. of SPIE*, Vol. 7935, No. 79350Y, pp. 1-12.

Tanaka, N.; Ueno, I.; Hiraga, T.; Tanigaki, N.; Mizokuro, T.; Yamamoto, N. & Mochizuki, H. (2007). Optically controlled optical-path-switching apparatus, and method of switching optical paths. *US patent*; 7,301,686.

Tanaka, N.; Ueno, I.; Hiraga, T.; Tanigaki, N.; Mizokuro, T.; Yamamoto, N. & Mochizuki, H. (2010). Optically controlled optical-path-switching-type data distribution appratus and distribution method. *US patent*; 7,792,398B2.

Ueno, I.; Tsujita, K.; Chen, G.; Tanaka, N.; Takarada, S.; Yanagimoto, H.; Moriwaki, D.; Mito, A.; Hiraga, T. & Moriya, T. (2003). Coaxial Configuration of the Gating and Signal Light for a Switching Device of a Dye-Dissolved Polymer Film. *Jpn. J. Appl. Phys.*, Vol. 42, No. 3, pp. 1272-1276.

Ueno, I.; Hiraga, T.; Mizokuro, T.; Yamamoto, N.; Mochizuki, H. & Tanaka, N. (2007). Optical path switching device and method. *US patent*; 7,215,491.

Ueno, I.; Tanigaki, N.; Yamamoto, N.; Mizokuro, T.; Hiraga, T.; Tanaka, N.; Nagaeda, H. & Shiga, N. (2010). Optical deflection method and optical deflection apparatus. *US patent*; 7,826,696B2.

6

Physical-Layer Attacks in Transparent Optical Networks

Marija Furdek and Nina Skorin-Kapov
University of Zagreb
Croatia

1. Introduction

For the past decades, network traffic has been showing immense growth trends, as we are witnessing the rapid development of network applications such as Internet Protocol TV (IPTV), peer-to-peer traffic, grid computing, multi-player gaming etc. Optical fiber, with its huge capacity of up to 50 THz, low bit error rate of 10^{-12}, low loss of 0.2 dB/km and low noise and interference characteristics has been widely accepted as a viable future-proof solution to meet the ever-increasing network bandwidth demands. In comparison with the available fiber capacity, the speed of edge electronic equipment of only a few Gb/s creates a bottleneck, so fiber bandwidth is divided into independent wavelength sets, each capable of carrying traffic between a pair of nodes at different speeds. This is the underlying principle of Wavelength Division Multiplexing (WDM), where different wavelengths supporting communication between different end users are multiplexed and carried simultaneously over the same physical fiber. Under normal operating conditions, carried wavelengths do not significantly interfere with each other inside the fiber. At the receiver's side, they are demultiplexed or filtered to ensure that every receiver receives the intended wavelength. An illustration of WDM principle is shown in figure 1.

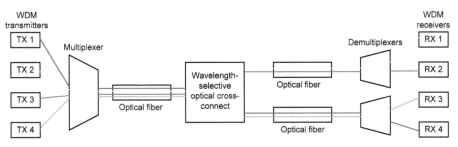

Fig. 1. Example of a simple WDM system.

In Transparent Optical Networks (TONs), signals do not undergo optical-electronical-optical (OEO) conversion at the intermediate nodes they traverse. Communication takes place entirely in the optical domain, via all-optical channels called lightpaths. The process of establishing lightpaths consists of finding a physical route and assigning a wavelength to each of them, called Routing and Wavelength Assignment (RWA). The set of established lightpaths then comprises a so-called virtual topology over the given physical topology. Intermediate nodes perform wavelength-switching without regenerating or even interpreting the carried signals. Namely, full 3R (re-amplification, re-shaping, re-timing) signal regeneration in the optical domain is still in the experimental phase. Therefore, optical signals can only be re-amplified (1R) in the optical domain, while re-shaping and re-timing require OEO conversion. We are currently witnessing the evolution of optical networking from opaque networks with all-electronic switching, implying OEO conversion at every node, to transparent networks with all-optical switching and no OEO conversions at intermediate nodes. Networks in which most of the nodes are transparent and some of them are strategically equipped with 2R and/or 3R regenerators to improve the quality of analog optical signals are called translucent (Shen & Tucker, 2007).

The absence of lightpath regeneration in transparent optical networks not only provides signal transparency to bit rates, protocols and modulation formats but also reduces the costs and energy consumption associated with OEO conversion. However, transparency introduces significant changes to the security paradigm of optical networks by allowing signals whose characteristics fall out of the protocol-specific bounds or component working ranges to propagate through the network undetected. This creates a security vulnerability which can be exploited by a malevolent user to perform deliberate attacks aimed at degrading the proper functioning of the network. Due to the high data rates and latency employed in back-bone optical networks, even sporadic attacks of short duration can cause large data and revenue losses.

Section 2 gives an overview of different types and methods of physical-layer attacks in TONs, along with experimental evaluation of some of the vulnerabilities of network components that can be exploited by malicious users. Section 3 gives an overview of the current issues and trends in attack management and control in TONs, as well as some methods and guidelines for increasing network resilience to attacks. Finally, Section 4 concludes this chapter.

2. TON vulnerabilities to physical-layer attacks

The high data rates employed by TONs make them extremely sensitive to communication failures, whether they result from component malfunctions caused by external factors or fatigue, or from deliberate attacks. However, the differences between component faults and deliberate attacks make their consequences and recovery scenarios fundamentally different. Namely, disruption caused by component faults is restricted to the connections passing through the affected component, so rerouting these connections using classical survivability mechanisms usually solves the problem until the component is replaced/fixed. On the other hand, attacks can propagate to many users and different parts of the network, significantly complicating their detection and localization. Furthermore, the traffic itself can be the source of attack so rerouting affected connections may even worsen the consequences of the attack, instead of alleviating them. Furthermore, attacks, unlike failures, may appear sporadically so as to avoid detection.

Overviews of various physical-layer attacks in TONs can be found in (Fok et al., 2011; Mas et al., 2005; Médard et al., 1997). An attacker can gain access to the physical network components as a legitimate user (or impersonating one) or by otherwise breaching into the network. The attacker may be an outsider or, equally likely, a person with inside access to the network facilities, according to (Richardson, 2008).

Depending on the intentions of the attacker, physical-layer attacks can be divided into two main groups:

a. Tapping attacks - aimed at gaining unauthorized access to data and using it for traffic analyses or eavesdropping purposes.
b. Service Disruption attacks - aimed at degrading the Quality of Service (QoS) or causing service denial.

Tapping attacks imply breaches in communication privacy and confidentiality. Occurrences of these attacks have been recorded in the past, e.g. in 2000 when three main trunk lines of the Deutsche Telekom network were breached at Frankfurt Airport in Germany or when an illegal eavesdropping device was discovered attached to Verizon's optical network in 2003 (Miller, 2007). The most likely purpose of these attacks was industrial espionage. Estimates indicate that only in the year 2000, corporate espionage cost US companies approximately $20 billion in purely technical means (Oyster Optics Inc., 2002).

The goal of service disruption attacks is to deteriorate the signal quality of legitimate communication channels. Depending on the severity of these attacks, their consequences may range from slight deterioration of the signal-to-noise ratio (SNR) to complete loss of service availability. They can also be aimed at manipulating communication by injecting false information or undermining the integrity of the transmitted data. Most commonly, these attacks are realized by injecting a malicious high-powered jamming signal which interferes with legitimate signals inside various network components. Methods of exploiting the vulnerabilities of the key building blocks of TONs (i.e. optical fibers, amplifiers and switches) to perform tapping and service disruption attacks are described in the following subsections.

2.1 Optical fibers

Optical fibers are immune to electromagnetic interference, which eliminates the possibility of eavesdropping through observation of side-channel effects, but, unless shielded, they are still susceptible to eavesdropping through other means. Namely, under normal operating conditions, light is kept inside the fiber core through total internal reflection, where the angle between the light beam and the core inner surface exceeds the critical angle and the beam is totally reflected back into the core. Bending the fiber violates the condition of total internal reflection of light inside the fiber core and causes part of the signal to be radiated out of the fiber, as shown in figure 2. If a photodetector is placed at the fiber bend, it can pick up such leakage and deliver the transmitted content to the intruder. Commercial tapping devices which introduce losses below 0.5 dB can be found on the market. There are also techniques which introduce losses below 0.1 dB, making such attacks extremely difficult to detect by network monitoring systems.

Some of these devices may cause a short interruption of service due to the necessity of cutting the fiber in order to install the device, after which the transmission is re-established. If this interruption is noticed, the technical personnel is quite likely to find the location of the tap, making this method short-lived (Witcher, 2005). However, some eavesdropping devices can be clamped onto the fiber and create micro bends causing leakage without actually cutting the fiber. Retrieval and interpretation of tapped data may require more sophisticated methods, depending on the signal wavelength, polarization, modulation format and other characteristics, but a well equipped attacker should be able to overcome these obstacles.

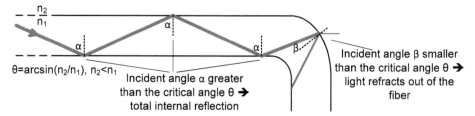

Fig. 2. Bending the fiber violates the conditions of total internal reflection and causes light to leak outside the fiber core.

Bending the fiber also enables a jamming signal to be inserted into the network. Under normal operating conditions, transmission effects in fibers are fairly linear, but high distances or high input powers increase the nonlinear effects among signals, of which four-wave mixing and cross-phase modulation are the most significant. A powerful jamming signal injected into the fiber enhances these effects and deteriorates the SNR of other signals. Due to the low attenuation of optical fibers, such a jamming signal can propagate from the entry point to other network components without losing its power and cause damage inside optical amplifiers and switches. This may be especially significant in new optical fiber access networks, where splitters and fibers are largely placed in public areas, with easy access to anyone.

2.2 Optical amplifiers

Erbium-doped fiber amplifiers (EDFAs) are the most commonly used optical amplifiers in today's WDM networks. They use an erbium-doped optical fiber core as gain medium to amplify optical signals. The energy of ionized erbium atoms can change between discrete levels. Atoms in lower energy levels have less energy and they can be raised to a higher level by absorbing an amount of energy equal to the difference between the two levels. Equivalently, a transition from a higher to a lower energy level results in the emission of a photon whose energy equals the difference between the two levels. In a normal state, the amount of erbium ions in the ground energy level is much higher than those in upper levels. To achieve amplification, the gain medium is pumped with an external source of energy which causes the number of ions in higher energy levels to exceed their number in lower levels, i.e. obtaining population inversion. When light of the appropriate frequency passes through such a medium, its photons stimulate the transition of excited electrons to lower energy levels, resulting in the stimulated emission of photons which have the same frequency, direction of propagation, phase and polarization as the incident photons. In this way, the incoming optical signal is amplified.

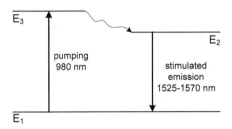

Fig. 3. Three energy levels of Er^{3+} ions in silica glass for 980 nm pumping. Each discreet energy level represents a continuous energy band.

Figure 3 shows three energy levels – E_1, E_2 and E_3 of Er^{3+} ions in silica glass. In reality, the energy levels shown here as discreet are spread into a continuous energy band. The energy difference between levels E_1 and E_3 corresponds to the energy of photons of light at 980 nm. When light at that wavelength is pumped into the erbium-doped fiber, its absorption causes the transition of ions from E_1 to E_3. Light at 1480 nm can also be used for pumping, but the pumping process is more efficient at 980 nm, resulting in a higher gain for the same pump power (Ramaswami & Sivarajan, 2002). The excited ions stay at the E_3 level for a very short time and then quickly transit to level E_2. The lifetime of the transition from level E_2 to E_1 is much longer, about 10 ms, and it is accompanied by the emission of photons on a wavelength between 1525 and 1570 nm. With pumping power high enough, the ions which fall back to level E_1 are quickly raised to E_3. The result of the synergy of these two processes is that most of the ions can be found at level E_2, i.e. population inversion between levels E_2 and E_1 is achieved. Under such conditions, light on wavelengths of 1525-1570 nm is amplified by stimulated emission from level E_2 to level E_1.

An optical amplifier is characterized by its gain, gain bandwidth, gain saturation, polarization sensitivity and amplifier noise (Mukherjee, 2006). The gain is defined as the ratio between the power of the signal at the output of the amplifier and its power at the input. Gain bandwidth specifies the frequency range over which the amplifier is effective. This parameter limits the number of wavelengths available in a network for a given channel spacing. Gain saturation is the value of output power after which an increase in input power no longer causes an increase in output power. It is usually defined as the output power at which there is a 3 dB reduction in the amplifier gain. Polarization sensitivity measures the difference in gain between two orthogonal polarizations of the dominant signal mode (HE$_{11}$ mode). The prevailing component of amplifier noise for EDFAs is Amplifier Spontaneous Emission (ASE), which arises from spontaneous transitions of ions from energy level E_2 to E_1, independent of any external radiation. Although the radiated photons have the same energy as the incoming optical signal, their frequency, phase, polarization and direction do not match.

EDFAs have several advantages over other types of optical amplifiers, such as Raman and semiconductor optical amplifiers. They provide high gain, are capable of simultaneous amplification of WDM signals independent of the light polarization state, have a low noise figure and low sensitivity to temperature (Laude, 2002). However, they also have drawbacks such as additional noise (ASE), dependency of gain on the spectrum and power of the incoming signal, and transients which occur when individual WDM channels are dropped.

If we consider each of the discrete energy levels in the doped fiber as a continuous energy band, then EDFAs are capable of simultaneously amplifying signals on several different wavelengths. As mentioned before, they most commonly amplify signals within the 1525-1570 nm wavelength range. However, due to the fact that the distribution of excited electrons is not uniform at various levels within a band, the gain of an EDFA depends on the wavelength of the incoming signal, with a peak around 1532 nm (Ramaswami & Sivarajan, 2002). This can be compensated for by employing passive or dynamic gain equalization (Bae et al., 2007; Laude, 2002). However, the limited number of available upper-state photons necessary for signal amplification must be divided among all incoming signals. Each of the signals is granted photons proportional to its power level, which can lead to so-called *gain competition*, where stronger incoming signals receive more gain, while weaker signals receive less. Due to the large number of input channels and high data rates employed in today's WDM networks, the dependency of EDFA gain assignment on the spectrum and power of the incoming signals can have a significant impact on network functioning.

Gain competition can be exploited to create service disruption as described in (Mas et al., 2005; Médard et al., 1998). In an *out-of-band jamming attack*, a malicious user injects a powerful signal (e.g. 20 dB above normal) on a wavelength different from those of other, legitimate signals, but still within the pass-band of the amplifier. The amplifier, unable to distinguish between the attacking signal and legitimate data signals, provides gain to each signal indiscriminately. The stronger, malicious signal will get more gain than the weaker, legitimate signals, robbing them of power. Thereby, the QoS level on the legitimate signals will deteriorate, potentially leading to service denial. Furthermore, the power of the attacking signal will have an additional increase downstream of the amplifier, allowing it to spread through other transparent nodes and affect other signals at their common EDFAs.

2.2.1 Laboratory assessment of gain competition

The impact of the jamming signal depends on its power level and wavelength. We tested this relation in laboratory setting (Furdek et al., 2010a) using two EXFO IQ-2600 tunable lasers sources, variable attenuators EXFO IQ-3100 to attenuate the signals and simulate losses in the optical fiber and an EDFA with 36 m of erbium-doped fiber Lucent Technologies HE-980 as the gain medium, pumped with a 980 nm pump signal from an Agilent FPL4812/C laser pump. One of the laser sources represented a legitimate signal with constant power (-25,51 dBm before entering the EDFA) and wavelength (1549,74 nm), while the other represented a powerful jamming signal with varying power and wavelength. Figure 4 shows the dependence of the amount of gain given to the legitimate signal on the power of the jamming signal on the next WDM channel, at 1549,05 nm. The power of the interfering signal was increased in 2 dB increments from the same level as the legitimate signal, until it was 20 dB stronger. The pump power was set to 40 mW. The measurements in figure 4 show how the amount of gain of the legitimate signal decreases in response to an increase in the power of the interfering signal. This is due to the fact that the interfering signal, as it becomes more powerful, consumes more and more upper-state photons in the EDFA, and thus robs the legitimate signal of gain.

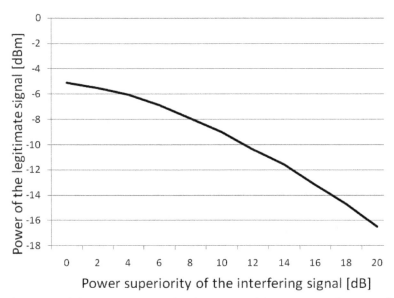

Fig. 4. The power of the legitimate signal at the output of the EDFA as a function of the power superiority of the interfering signal on the neighboring WDM channel, at 40 mW EDFA pumping power.

The gain of the legitimate signal also depends on variations in the wavelength of the interfering signal of constant, high power. Figure 5 shows this dependency for the legitimate signal at 1549,74 nm and a 20 dB stronger interfering signal whose nominal wavelength varies from 1530 nm to 1550 nm, in 5 nm increments. The influence of different operating points of the EDFA on the output power of the legitimate signal in this scenario was investigated by changing the pump power from 40 mW to 80 mW. In figure 5, *P_legit* denotes the power of the legitimate signal, and *P_interfering* the power of the interfering signal at the EDFA output. Power levels measured for pump powers of 40 mW and 80 mW have suffixes *_40mWpump* and *_80mWpump*, respectively. From figure 5, it can be seen that the amount of gain robbed from the legitimate signal by the high-powered jamming signal increases as their wavelength separation decreases.

Table 1 summarizes the influence of wavelength separation and power superiority of the interfering signal over the legitimate signal at 1549,74 nm. In the first case, the wavelength of the interfering signal matches the used EDFA gain peak at 1531 nm. In the second case, it is at the first neighboring WDM channel, i.e. at 1549,08 nm. For both cases, we investigate the gain of the legitimate signal for jamming signal power levels 10 dB and 20 dB higher than the legitimate signal. For two pump powers, i.e. 40 mW and 80 mW, the first row in the table shows the gain of the legitimate signal when no jamming signal is present. The values in the table clearly show that the presence of a strong signal results in weaker amplification of the signal at lower power level. The gain of the legitimate signal drops as the power of the interfering signal increases. Furthermore, for a given power level of the interfering signal, its harmful effect to the legitimate signal is more intense when their wavelengths are close in the spectrum, as highlighted in the table.

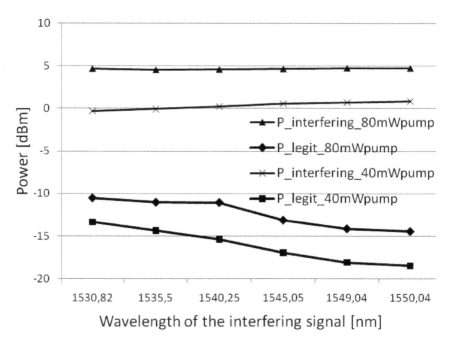

Fig. 5. The power of the legitimate and interfering signal at the output of the EDFA as a function of the wavelength of the interfering signal, at 40 mW and 80 mW EDFA pumping power and the interfering signal 20 dB stronger than the legitimate.

Out-of-band jamming can also be used to tap a signal. In some optical amplifiers, gain competition occurs at the modulation rate, which enables tapping by observing cross-modulation effects.

EDFA pump power [mW]	Power superiority of the interfering signal [dB]	Wavelength of the interfering signal [nm]	Gain of the legitimate signal [dB]
40	-	-	20,34
	10	1530,84	17,51
		1549,08	**15,38**
	20	1530,84	12,62
		1549,08	**8,01**
80	-	-	22,68
	10	1530,84	20,61
		1549,08	**20,03**
	20	1530,84	15,63
		1549,08	**11,59**

Table 1. An overview of the gain of the legitimate signal at 1549,74 nm for different test scenarios, with the power of the interfering signal at 10 and 20 dB above that of legitimate signal.

2.2.2 Amplifier cascades

When EDFAs are used in a cascade, the flatness of their gain becomes a critical issue. Namely, slight differences between the amounts of gain available for signals at different wavelengths get multiplied as they traverse the cascaded amplifiers. Because of this, signals on certain wavelengths might get amplified several times, while others may suffer from significant SNR deterioration (Ramaswami & Sivarajan, 2002). This situation is shown in figure 6. There are several ways of dealing with this issue. For example, signals on different wavelengths can be pre-equalized, so that the signals on wavelengths with higher gain are attenuated, and those with lower gain are amplified before entering the cascaded amplifier segment. Another way of dealing with the problem is to introduce gain equalization at each amplifier stage.

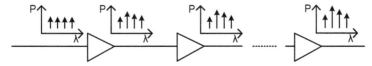

Fig. 6. The cumulative effect of unequal amplifier gain at different wavelengths after a cascade of amplifiers.

In case of cascaded EDFAs, power transients potentially present a great security threat. Due to the fact that the amplifier gain depends on the total input power, the failure of one channel will lead to surviving channels getting more gain and, thus, have higher power when they arrive to their receivers. This means that setting up or tearing a channel down affects other channels that share amplifiers with it (Karásek & Vallés, 1998). This effect may cause serious problems in dynamic optical networks where suppression of transients becomes increasingly important. A typical amplifier implementation used in today's networks consists of two EDFA stages working in gain mode, where setting up a new channel will not affect power levels of existing channels (Zsigmond, 2011). Automatic gain control (AGC) solves the problem of transients by monitoring the power levels in different ways and keeping the output power per channel constant, regardless of the input power. In such a network, high-power signals could not propagate. However, this is only valid for deviations of power within a certain window defined by the component specifications. If the difference between the power of the jamming signal and the normal users' signals exceeds this range, amplifiers with AGC may not be able to provide power equalization. (Way et al., 1993) proposed optical limiting amplifiers able to limit the output power of all signals within a dynamic range of input power and thwart the propagation of jamming attacks, but at a trade-off with a higher price of such equipment. Today, most commercially available amplifiers are capable of monitoring channel power and reducing the excessive power levels of jamming signals (Zsigmond, 2011). However, (Deng & Subramaniam, 2004) describe an attack which can affect even networks with ability to equalize excessive power levels. It is referred to as a *low power QoS attack*. Amplifier placement along the link usually ensures compensation for the preceding fiber span. If an attacker attacks a splitter at the beginning of a link, they are able to attenuate the power of the signal more than the amplifier is able to compensate for. Such induced attenuation can significantly degrade the performance metrics of attacked lightpaths. The attenuation at the end of the link on which the splitter is installed may not be significant enough to generate an alarm at that exact location, but it may cause other network elements with power equalization capabilities (e.g., switches) to reduce the power of other signals in an effort to maintain an even distribution of power among channels. Hence, other lightpaths suffer from attenuation and may cause the same effect in other parts of the network. When service degradation along a lightpath finally crosses the preset threshold, the location of the raised alarm may be far from the original placement of the attached splitter. This type of an attack may be especially significant for networks employing Raman amplifiers, whose usage is increasing in long haul transmission suffering from high attenuation (Zsigmond, 2011). Security advantages of Raman amplifiers include more reliable amplification, higher saturation power than EDFA and more accurate monitoring, resulting in faster generation of alarms in case of signal anomalies (Islam, 2003). However, output powers of Raman amplifiers are high and require splicing. Multiple splices can cause the Raman pumps to be reflected and, thus, highly reduce the amplifier gain. This vulnerability can be a target of a planned attack, possibly leading to a link outage (Zsigmond, 2011). Furthermore, Raman amplifiers require high-power pump sources at the right wavelength and an attacker with inside access to an amplifier may endanger the amplification process by tampering with any of these parameters.

2.3 Optical switches

The main functions of wavelength-selective optical cross-connects (OXC), also referred to as reconfigurable wavelength routing switches, can include lightpath provisioning, wavelength switching, protection switching (rerouting connections), wavelength conversion and performance monitoring. Such optical switches usually consist of demultiplexers, photonic switching fabric and multiplexers. A typical architecture of a wavelength-selective OXC is shown in figure 7.

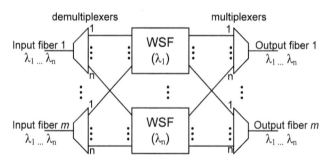

Fig. 7. The typical architecture of a wavelength-selective OXC, consisting of multiplexers, demultiplexers and wavelength switching fabric (WSF).

The incoming signal is first decomposed by demultiplexers into constituent wavelengths, which are then directed each onto their own switching fabric. Multiplexing and demultiplexing can be realized using Arrayed Waveguide Gratings (AWGs), Thin-Film Filters (TFF), Mach-Zehnder Interferometers (MZIs), Fiber Bragg Gratings (FBG) and other. The Wavelength Switching Fabric (WSF), i.e., the central part of the node, performs transparent switching of WDM channels from their input to output ports. Optical switches can be implemented using 2D or 3D Micro-Electro-Mechanical Systems (MEMS), semiconductor optical amplifier (SOA) gates, holographic switches, liquid crystal, and thermo-optical or electro-optical technologies (Tzanakaki et al., 2004). The WSF can be reconfigurable or fixed. A fixed or non-reconfigurable switching fabric has manually hard-wired connections between input and output ports, which cannot be changed on demand. On the other hand, connections between input and output ports of reconfigurable WSFs can be dynamically reconfigured in times ranging from several milliseconds (MEMS, bubble, liquid crystal, opto-mechanical, thermo-optic switch), several microseconds (acousto-optic switch) to several nanoseconds (electro-optic, SOA-based switch) (Papadimitriou et al., 2003; Rohit et al., 2011). After switching is performed, wavelengths intended to each output fiber are combined by multiplexers.

The main security vulnerability of optical switches arises from their proneness to signal leaking, giving rise to crosstalk. Almost all TON components, i.e., filters, multiplexers, demultiplexers and switches, introduce crosstalk in one form or another. Malicious users can take advantage of this phenomenon to cause service degradation and/or perform eavesdropping.

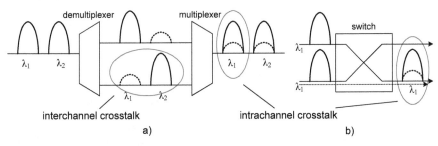

Fig. 8. (a) An optical multiplexer/demultiplexer and (b) an optical switch as sources of interchannel and intrachannel crosstalk.

In general, there are two types of crosstalk in transparent optical networks – interchannel and intrachannel crosstalk. Interchannel crosstalk occurs between signals on sufficiently spaced wavelengths, i.e. such that they do not fall inside each other's receiver pass-bands. Adjacent channels are usually the primary sources of crosstalk, while the influence of channels with higher wavelength separation is usually negligible. Inside OXCs, this type of crosstalk arises from non-ideal demultiplexing, where one channel is selected while the others are not perfectly dropped. This scenario is shown in figure 8(a). Depending on the implementation of the (de)multiplexers, their levels of crosstalk may range from 12 dB for TFF to 30 dB for AWG, MZI and FBG (Mukherjee, 2002). Intrachannel crosstalk occurs among signals on the same wavelength, or signals whose wavelengths fall within each other's receiver pass-band.

Multiplexers, demultiplexers and optical switches can all be sources of intrachannel crosstalk. Namely, when demultiplexers separate incoming signals at different wavelengths, a small portion of each signal leaks onto ports reserved for signals at other wavelengths. After switching, when multiple signals at different wavelengths are multiplexed back onto the same output fiber, small portions of a certain wavelength that had leaked onto other wavelengths can leak back onto the common fiber (Rejeb et al., 2006b). Consequently, the signal on that wavelength will have crosstalk originating from its very own components carrying the same information, but suffering from different delays and phase shifts, as shown in figure 8(a). Intrachannel crosstalk can also arise in optical switches due to non-ideal switching. Namely, switching ports are not perfectly isolated from each other, so components of different signals transmitted on the same wavelength can leak and interfere with each other. Since the damaging signal is on the same wavelength as the legitimate signal, intrachannel crosstalk cannot be filtered out by optical filters or removed by demultiplexers (Deng et al., 2004). Figure 8(b) shows an optical switch as a source of intrachannel crosstalk. Crosstalk levels of optical switches range from 35 dB (SOA, liquid crystal, electro-optical, thermo-optical and holographic switches) to 55 dB for MEMS.

Optical couplers are the basic building blocks of optical switches, multiplexers and demultiplexers, modulators, filters and wavelength converters (Ramaswami & Sivarajan, 2002) and are the source of a significant amount of inter/intra-channel crosstalk. Generally, an optical coupler is a device used to combine or split signals in an optical network and can be passive or active. In passive couplers, employed in TONs, signals are redistributed without opto-electrical conversion and do not require any external power.

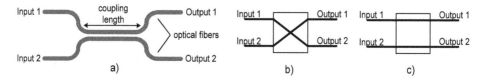

Fig. 9. A (a) directional coupler and its two states: (b) cross state and (c) bar state.

A passive directional 2×2 coupler is shown in figure 9(a). It consists of a pair of parallel optical waveguides in close proximity. The most commonly used couplers, called fused fiber couplers, are made by fusing two fibers together in the middle (Ramaswami & Sivarajan, 2002). The fraction of the signal power that is transferred from the input to the output of an optical waveguide is defined by the coupling ratio α, denoting that a fraction α of the power of the signal at the input of a waveguide is transferred to its output, while the remaining $1-\alpha$ of the power is directed to the output of the other waveguide. Ideally, all the input power on one waveguide of a directional coupler is coupled to the other waveguide for the cross state, while in the bar state there should be no coupling between the two waveguides.

Figures 9(b) and (c) show the cross state and the bar state of an optical coupler, respectively. In reality, however, light is not perfectly coupled and components of signals from different waveguides leak onto unintended outputs, giving rise to crosstalk. Non-ideal signal coupling also causes signal losses and attenuation, which can be compensated by placing optical amplifiers at the splice output. In this way, however, the desired part of the signal will be amplified as well as the undesired part, which makes crosstalk the main deficiency of optical couplers (Vaez & Lea, 2000). Crosstalk in a directional coupler is defined as the ratio of light power at the undesired output port to the power at the desired output port with crosstalk levels varying between -20 dB and -30 dB. It can occur for various reasons, including waveguide asymmetry, absorption loss, non-optimal coupling length, unequal excitation of the symmetric and asymmetric modes at the input, or fabrication variations (Chinni et al., 1995).

Couplers can be wavelength selective, and they are often used to combine signals at 1310 nm and 1550 nm onto a single fiber, or to split them from the same incoming fiber to two different outputs. In the latter case, due to crosstalk, small portions of the signal passing through the coupler are directed onto unintended outputs, deteriorating the Signal to Noise Ratio (SNR) of the signal which was intended for that output. Levels of this crosstalk depend on the exact wavelengths of the incoming signals.

2.3.1 Laboratory assessment of crosstalk in optical couplers

We tested the crosstalk of couplers in a laboratory setting from (Furdek et al., 2010b), using a FIS WDM13500129U coupler/splitter with SMF28 Singlemode fiber, operating at wavelengths 1310/1550 nm +/- 20 nm. This coupler was used as a wavelength-selective splitter for dividing the incoming WDM signal from the input port into its constituent wavelengths to two different output ports, i.e. one for signals at 1310 nm, and the other one for 1550 nm.

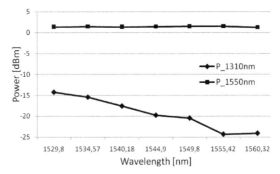

Fig. 10. Power at the coupler outputs dedicated to wavelengths at 1310 and 1550 nm for different wavelength of the incoming signal.

Figure 10 shows the effects of imperfect splitting of the incoming signal to ports dedicated to wavelengths at 1310 and 1550 nm, i.e. the power of the incoming signal at various wavelengths near 1550 nm present at the 1310 nm output. As the wavelength of the incoming signal decreases from 1560,32 nm to 1529,90 nm (in 5 nm steps), and approaches the central frequency of the 1310 nm output, the undesirable leakage intensifies.

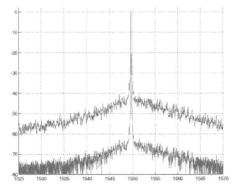

Fig. 11. The spectrum of the incoming signal at 1550 nm on the output port corresponding to 1550 nm (upper line) and on the output port corresponding to 1310 nm (lower line).

Figure 11 shows this effect for incoming signal at a nominal wavelength of 1550 nm. The upper line shows the spectrum of the signal at the output port corresponding to wavelengths around 1550 nm, while the lower line shows the spectrum at the output port corresponding to wavelengths around 1310 nm. The peak of the signal recorded at the 1310 nm-output clearly indicates the amount of the signal at 1550 nm that had leaked onto the unintended output. The signal power level of 1,48 dBm at the 1550 nm port, combined with -20,50 dBm at the 1310 nm port, indicates that the level of crosstalk is -21,98 dB. This value by itself is not large enough to significantly impact signal quality. However, many network components consist of several cascaded optical couplers, which all contribute to the overall level of crosstalk. Furthermore, signals traverse numerous components on their path from source to destination. When these factors combine, enough crosstalk can accumulate over the propagation path of a signal for the risk of service degradation to increase significantly even in cases when there is no high-powered jamming signal. When such a signal is present in the network, it causes an additional increase in the leakage inside couplers and components they comprise, resulting in a significant damage to co-propagating user signals.

2.3.2 Crosstalk attacks

Although crosstalk originating from direct couplers can have a significant impact on the overall Quality of Service (QoS) in the network, problems caused by crosstalk in optical networks can go beyond such signal quality deterioration. Namely, a malevolent user can take advantage of crosstalk to perform attacks aimed at eavesdropping, tapping, and/or degrading the quality of service (QoS) of other users. An overview of methods using crosstalk for attack purposes can be found in (Mas et al., 2005).

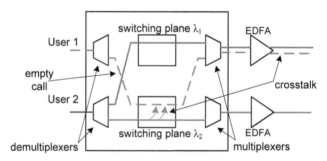

Fig. 12. An example of a tapping attack exploiting intra-channel crosstalk in a wavelength-selective switch.

Figure 12 shows an example of a tapping attack exploiting intrachannel crosstalk in a wavelength-selective switch, as described in (Médard et al., 1997). The upper input port is not used, while the bottom port receives incoming signals on wavelengths λ_1 and λ_2. Each of the signals on those two wavelengths is switched on its own switching fabric. Due to mechanisms of intrachannel crosstalk in demultiplexers, multiplexers and switching fabric described in the previous sections, components of both signals leak onto unintended outputs and get amplified by the power amplifier (EDFA). If a tapper gains access to one of the unused output ports, e.g. the upper output port in figure 12, part of the signal at λ_2 is delivered straight into his hands. This problem can be solved by individually amplifying only signals on connections which are registered at the network management system. However, an attacker can still request a legitimate data channel and then not send any information over it, but use it to tap other signals at the same wavelength. In figure 12, the tapper is User 1, whose false data connection at wavelength λ_2 picks up components of User 2's legitimate connection at the same wavelength that had leaked inside their common switch.

Intrachannel crosstalk enables *in-band jamming*, an attack method in which an attacker inserts a powerful signal within the signal window of the legitimate user he is trying to affect. Consequently, two signals may undesirably exchange information at their common switch.

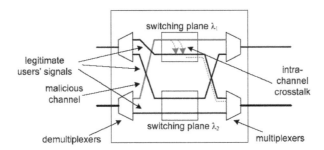

Fig. 13. An example of a jamming attack exploiting intra-channel crosstalk in an optical switch.

Figure 13 shows an example of a jamming attack via intrachannel crosstalk in optical switches. Here, an attacker injects a high-powered signal on the same wavelength as other, legitimate data signals. Components of the high-powered signal will leak onto adjacent channels inside their common optical switches, impairing the quality of the transmission on those signals. If the attacking signal is strong enough, it is possible that enough power is transferred onto adjacent channels inside their common switch, for them to gain attacking capabilities. Consequently, the attacked signal becomes an attacker itself, allowing the attack to propagate through the network, affecting signals which do not even share any physical components with the original attacking signal. This type of attack is shown in figure 14. Via intra-channel crosstalk in switches, the attacker managed to affect not only user 1's legitimate signal, but the attack also propagated to users 2 and 3, which share no common physical components with the original attacker. This type of attack is particularly hazardous to network operation since the nature of its propagation makes localization of the original source of attack very difficult.

Jamming attack exploiting intrachannel crosstalk in switches has been previously identified in the literature by (Wu & Somani, 2005), and recently (Peng et al., 2011) provided an experimental validation of the proposed attack model. They proved through simulation that high-power jamming attacks indeed possess propagation capabilities in affecting other lightpaths at the same wavelength via intrachannel crosstalk inside their common switches and lightpaths at different wavelengths via interchannel crosstalk inside their common fibers. The propagation of intrachannel crosstalk attacks ends after at most three stages of optical switches, while interchannel crosstalk attacks get attenuated after traversing three fiber segments. This means that, in the scenario from figure 14, the signal quality of user 3 would not suffer from serious BER degradation from the attacker's jamming signal.

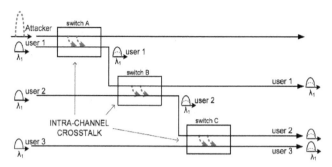

Fig. 14. Propagation of intra-channel crosstalk attacks in an all-optical network.

The vulnerability of TONs to high-power jamming attacks depends on employed hardware components and node architectures, as well as the architecture of the established virtual topology. Besides the wavelength-selective (WS) realization of OXCs, they can also be implemented as broadcast-and-select (B&S) devices. In B&S architecture, the wavelength switching fabric is replaced by passive splitters and couplers which connect the incoming signal to tunable filters. After filtering the desired wavelength, the signals from filter outputs are coupled onto the desired OXC output port. (Arbués et al., 2007) report that B&S architectures exhibit greater vulnerability to in-band jamming due to low isolation of tunable filters. WS architecture performs slightly better due to improved isolation at the multiplexing and demultiplexing stages.

(Liu & Ji, 2007) studied the impact of the physical topology in conjunction with its constituent devices and the network traffic on the network resilience to in-band jamming attacks. Under the assumption of a fully connected virtual topology, i.e. a connection between each node pair and assuming that jamming attack propagation is not possible, they find that fully-connected mesh, star and ring physical topologies are the least resilient to attacks. The main cause of low resilience of a fully-connected mesh is its high nodal degree and, hence, a high expected number of affected channels at each node. The latter is also the reason why star networks are highly susceptible to attacks. For ring topologies, their vulnerability stems from large route lengths. A chord network topology is distinguished as the most resilient to attacks, with a logarithmic increase in resilience loss for a linear growth of the network size.

3. Security in TONs

As previously mentioned, the high data rates and huge throughput associated with transparent optical networks make them extremely sensitive to communication failures caused by component faults or deliberate attacks. A secure network should provide physical security of communication, i.e. provide service availability, guarantee a certain level of QoS and protect data integrity and privacy of communication. It should also ensure semantic security, i.e. protect the confidentiality and the meaning of data through authentication and cryptography mechanisms. Transparent optical transmission and the properties of attacks as described in the previous section impose a new set of demands on the network management system (NMS), responsible for network configuration, performance engineering, fault handling and the secure and safe functioning of the network (Rejeb et al,. 2010; Li et al., 2002).

The headstone of an efficient NMS in TONs is a flexible and robust control plane, which relies on accurate and timely monitoring in the optical domain. Control plane functions can roughly be divided into following tasks (Rejeb et al., 2010; Saha et al., 2003):

- Resource management – accurate information on resource availability must be available at all times and updated upon lightpath establishment or tear-down.
- Lightpath provisioning – initially, the topology and resources must be automatically discovered. For each incoming lightpath demand, the control plane should calculate a physical route based on the available resources and tentative QoS requirements. For this, accurate information regarding resource availability and the associated service quality is crucial.
- Signaling – information exchange regarding connection establishment, maintenance and tear-down between nodes, as well as the management of alarms in cases of failures, must be present.

Optical network security requires protective and/or preventive measures which minimize network accessibility to attackers, limit attack propagation and reduce the damage proportions inflicted by attacks. However, when an attack occurs in spite of these mechanisms, the NMS needs to undertake the following steps:

- Detect the attack – discover a deterioration of signal quality, an intrusion in the fiber, a loss of service or any other direct consequence of an attack. After detecting the presence of an attack, its exact location must be determined and the source of the attack must be identified.
- React to the attack – by triggering reaction mechanisms, the attacker's access point must be isolated and the harmful effects must be neutralized. The affected connections must be restored and communication should resume as fast as possible.

3.1 Protection and prevention of attacks

The risks and damage associated with physical-layer attacks can be alleviated through careful network planning, employment of additional equipment, quick and accurate post-attack recovery and optical cryptography. Achieving complete protection requires large investments by the network operator and may be economically unviable. Thus, an advantageous trade-off between the costs and achieved protection must be found. Attack protection may include the following measures (Fok et al., 2011; Médard et al., 1997):

- Hardware measures – shielding the fiber to protect from tapping, introducing additional equipment in the network capable of limiting excessive power (e.g. optical limiting amplifiers or variable optical attenuators), or using optical fuses which melt under high power (Shuto et al., 2004) to protect from high-power jamming. Using components with lower crosstalk levels also helps reduce the risk from jamming and tapping attacks.
- Transmission schemes – applying different modulation and coding techniques or limiting the bandwidth and power of certain signals may help against tapping and jamming.
- Architecture and protocol design – identifying and avoiding risky links or assigning different routes and wavelengths to separate trusted from untrusted users may decrease the risk. Here, assessment of link risk and user trustworthiness is crucial, which may be extremely complicated.

- Optical encryption – protects communication confidentiality by making it incomprehensible to an eavesdropper.
- Optical steganography – protects communication privacy by hiding the transmission between a pair of users underneath the public transmission channel. In this way, an attacker is unaware of the existence of communication, which makes it extremely difficult to perform tapping or jamming. However, the overall network vulnerability to jamming attacks may result in hidden communication being a collateral victim of jamming public channels.
- Optical network survivability – intelligent protection schemes can increase resilience to attacks by switching the signals under attack to unaffected parts of the spectrum or to physically disjoint backup paths.

Prevention may play a significant role in enhancing TON resilience to attacks, as well as the reduction of the deteriorating effects of attacks. The concept of attack-aware optical networks planning to reduce attack consequences was introduced in (Skorin-Kapov et al., 2010). By determining the mutual jamming attack relations between lightpaths, a novel objective criterion for the routing and wavelength assignment problem was defined, called the Lightpath Attack Radius (LAR). By minimizing the LAR of each lightpath through judicious routing, the maximum possible damage caused by such attacks can be reduced. In (Furdek et al., 2010c), a similar approach was developed for minimizing crosstalk effects caused by in-band jamming through judicious wavelength assignment. Our current ongoing work in attack-aware optical network planning is focused on survivability mechanisms and node power equalization placement.

3.2 Attack detection

Detection of an attack relies closely on reliable and accurate monitoring methods. In TONs, real-time monitoring must take place in the optical domain, without electronically interpreting the carried data. Descriptions of techniques for monitoring various optical signal parameters can be found in (Ho & Chen, 2009; Kilper et al., 2004). Depending on the technology, monitoring methods should be capable of measuring parameters such as channel power (peak and average) and aggregate WDM signal power, eye diagram, optical spectrum, polarization state, phase, pulse shape, Q-factor, chromatic and polarization-mode dispersion (PMD) etc. The measured parameters indicate the level of quality of aggregate WDM layer parameters, as well as individual signal quality parameters. Due to high prices of monitoring equipment, placing their minimal number in strategic locations and establishing supervisory channels able to detect as many faults as possible remains an important network planning problem. Today, there are commercially available reconfigurable optical switches which provide per-channel power and wavelength monitoring, such as that from (Cisco, 2011). Furthermore, they are usually equipped with variable optical attenuators and are, thus, able to dynamically react to excessive power levels on individual channels and thwart jamming attacks. However, these devices are not yet widely deployed. Currently, around 80% of deployed network nodes consist of fixed optical switches and add-drop multiplexers (FOADMs) whose power settings are determined in the system commissioning phase and do not offer the capability of dynamically managing power level fluctuations of incoming signals. Current market trends show a tendency of reconfigurable node usage increasing to 50% of network nodes in the next few years, while the remaining nodes will still consist of FOADMs (Zsigmond, 2011).

Some monitoring methods which can detect specific attack scenarios are elaborated in (Médard et al., 1997). These methods can rely on statistical analysis of the optical properties of transmitted signals or they can use special, dedicated signals. Statistical methods include wideband power detection and optical spectrum analysis. The first method measures the power over a wide bandwidth and reacts to deviations from statistically computed expected power levels. It may be able to detect a high-powered in-band jamming attack, but sporadic jamming attacks, jamming attacks which deteriorate the SNR without changing the power levels in the affected signals or tapping attacks which tap a very small amount of the total signal power may not be detectable by this method. The second method, i.e. optical spectrum analysis, measures the shape of optical signal spectrum. It is able to detect an out-of-band jamming attack causing gain competition, but in-band jamming may go undetected if the attacking signal doesn't introduce significant spectrum changes. This method isn't very helpful at detecting tapping attacks, unless the analyzer is placed on the link which drains the tapped portion of the signal and under the condition that it is able to distinguish authorized from unauthorized communication.

Two of the most common monitoring methods which use dedicated signals are the pilot tone method and optical time domain reflectometry (OTDR). Pilot tones are special signals dedicated to detecting transmission interruption. They may be carried along the legitimate signal's path at a different frequency. Their application in detecting in-band jamming requires very complex scenarios, because a pilot tone can only detect jamming on the very same frequency. Furthermore, pilot tones may be jammed themselves, creating an opportunity to mask jamming on legitimate lightpaths. Gain competition attacks may be discovered by pilot tones, but only if they receive amplification from the same EDFA as affected lightpaths. Even in this case, the BER degradation of the pilot tone caused by gain competition may go undetected because their main purpose is only to assert availability of communication, and not the available QoS. Pilot tones provide little help in detection of tapping, which would require a significant degradation of the signal quality. The main principle of OTDR is to inject pilot tones onto a link and analyze its echo in order to determine fiber cuts or losses, which makes attack detection abilities of these two methods similar. Detection of in-band jamming differs from the pilot-tone method only in its occurrence at the front-end of the link. Due to the fact that EDFAs are unidirectional, the OTDR method will not be able to detect gain competition. On the other hand, it may be successful in detecting tapping, which causes discontinuities in the reflected pilot tone.

3.3 Reaction to attacks

Once the presence of an attack is detected in the network, the NMS will try to eliminate it as soon as possible and re-establish reliable communication. Reaction from an attack at the optical layer should be fast and recovery should take place before the upper, slower network layers activate their reaction mechanisms. In most cases, the link on which the presence of an attack was detected will be switched off, which will trigger mechanisms for network survivability. Survivability mechanisms include protection, where resources are reserved for pre-computed backup paths of each of the working paths at lightpath setup time, and restoration, in which backup paths are computed upon a failure of the working path. Protection can be dedicated, where each backup path has its own dedicated resources, or shared, where resource sharing among backup paths of link-disjoint working paths is allowed. After finding a backup path for the affected connections, transmission will resume. Finding the exact location of the attack and disabling the attacker before re-establishing transmission of affected connections is crucial for this step. If these conditions are not met, protection resources may be wasted and switching the transmission to backup paths may even enhance attack propagation and worsen its effects.

A standardized approach for attack management has not yet been established. The main reason for this is the fact that optical monitoring technology hasn't yet reached its maturity and cannot provide reliable attack detection (Rejeb et al., 2006b), as well as the fact that the fault and localization methods design highly depends on the specific physical layer details (Rejeb et al., 2006a). Several frameworks for managing physical-layer attacks have been proposed in the literature. Reliable attack detection in some of them is based on the currently unrealistic assumption that all nodes are able to provide per channel monitoring, while others propose efficient monitoring placement policies, matching more realistic network scenarios.

Initial works on attack source identification date from the late 90's. In (Bergman et al., 1998), the authors propose a distributed algorithm for localizing jamming attacks based on the relation between the signal power metrics at the output and input of each node. Neighboring nodes exchange messages and determine the presence of an attack. The nodes are aware of their positions along every connection (i.e., whether they are upstream or downstream from the neighboring node they exchange messages with) so the algorithm is able to find the most upstream node which detects an attack along a connection, and thus can identify the source of the attack.

In the next decade, (Wu & Somani, 2005) provide a model of jamming attacks exploiting intrachannel crosstalk in optical switches with propagation capabilities, which enable affected lightpaths to acquire attacking capabilities and spread the attack to lightpaths which do not share any common physical components with the original attacker. They identify the assumption of all nodes being able to monitor all channels as unrealistic due to the high costs of this solution and propose a monitoring node model, their sparse placement, an additional test connection setup policy and a lightpath routing policy which is able to localize the source of a single crosstalk attack in the network.

In (Mas et al., 2005), the problem of finding the exact location of the failure is extended to the presence of single and multiple failures in cases where alarms can be false and/or lost. This problem is NP-complete even when no false or lost alarms exist. The algorithm is based on building a binary tree whose branches correspond to sets of network elements which will raise an alarm when a particular network component fails. Alarms differ according to the type of the failure and equipment used. When alarms are raised during network operation, the location of the failure is determined by traversing the binary tree and finding the components whose corresponding failures justify the received alarms. The authors also propose an optimal monitoring placement scheme for minimizing the number of network elements which are candidates to have a failure and, thus, minimizing the result given by the failure location algorithm.

(Rejeb et al., 2006a) investigate the local correlation of security failures and attacks at each OXC node and mechanisms to discover the tracks of multiple attacks through the network using as little monitoring information as possible. The correct functioning of this distributed algorithm relies on a reliable NMS which provides correct message passing and processing at local nodes. Namely, the algorithm uses updated connection and monitoring information at the input and output sides of any OXC node in the network. In order to decrease these tight requirements on monitoring information, the health of lightpaths which simultaneously propagate through OXC nodes is estimated through correlation with other lightpaths. When a node detects serious performance degradation along a lightpath at its output side, it runs a generic procedure for localizing the set of lightpaths which traverse this node and are most likely to be the offender. The localization procedure is then delegated to the next upstream node along each of these lightpaths which also registers performance degradation, and this is repeated until no such node is found.

In (Stanic & Subramaniam, 2011), the authors propose a fault localization scheme which collects monitoring information from lightpaths which carry traffic and from additionally established supervisory lightpath, achieving complete fault localization coverage. The authors consider a monitoring model where each OXC node is capable of detecting in-band loss–of–light faults. The problem of deciding which supervisory lightpaths will be added to the given set of traffic lightpaths is formulated as an Integer Linear Program (ILP) and an efficient heuristic approach for computing the optimal set of supervisory lightpaths is proposed.

4. Conclusion

This chapter presents an overview of the vulnerabilities of Transparent Optical Networks (TONs) to various physical-layer attacks. Furthermore, methods for attack detection and localization, as well as various countermeasures against attacks are described. As a result of the vulnerabilities associated with TONs stemming from optical components, transparency and high speed, new approaches to network security are increasingly needed as networks migrate to all-optical transmission. Such security frameworks require new, tailored attack detection, localization and network restoration mechanisms. In addition to upgrading existing ways of dealing with network failures and attacks, significant attention should be paid to prevention mechanisms, attack-aware planning and improved optical monitoring methods.

5. Acknowledgements

This work was supported by projects "A Security Planning Framework for Optical Networks (SAFE)", funded by the Unity Through Knowledge Fund (UKF) in Croatia, and 036-0362027-1641, funded by the Ministry of Science, Education and Sports, Croatia.

6. References

Arbués, P.G., Mas Machuca, C. & Tzanakaki, A. (2007). Comparative Study of Existing OADM and OXC Architectures and Technologies from the Failure Behavior

Perspective. *Journal Of Optical Networking*, Vol. 6, No. 2, (February 2007), pp. (123-133), ISSN 1536-5379

Bae, J.K., Koh, D., Kim, S.H., Park, N. & Lee, S.B. (2007). Automatic EDFA Gain Spectrum Equalization Using LPFGs on Divided Coil Heaters, *Proceedings of Optical Fiber Communication and the National Fiber Optic Engineers Conference* (OFC/NFOEC), ISBN 1-55752-831-4, Anaheim, USA, March 2007

Chinni, V.R., Huang, T.C, Wai, P.-K.A., Menyuk, C.R. & Simmonis, G.J. (1995). Crosstalk in a Lossy Directional Coupler Switch, *Journal of Lightwave Technology*, Vol.13, No. 7, (July 1995), pp. (1530-1535), ISSN 0733-8724

Cisco (2011.) Cisco ONS 15454 Multiservice Transport Platform. Available from <http://www.cisco.com/en/US/prod/collateral/optical/ps5724/ps2006/ps5320/product_data_sheet09186a00801849e7.html>

Deng, T. & Subramaniam, S. (2004). Covert Low-Power QoS Attack in All-Optical Wavelength Routed Networks, *Proceedings of IEEE GLOBECOM '04*, ISBN 0-7803-8595-3, Dallas, USA, November 2004

Deng, T., Subramaniam, S. & Xu, J. (2004). Crosstalk-aware wavelength assignment in dynamic wavelength-routed optical networks, *Proceedings of BroadNets'04*, ISBN 0-7695-2221-1, San Jose, USA, December 2004

Fok, M.P., Wang, Z., Deng, Y. & Prucnal, P.R. (2011). Optical Layer Security in Fiber-Optic Networks. *IEEE Transactions on Information Forensics and Security*, Vol. 6, No. 3, (September 2011), pp. (725-736), ISSN 1556-6013

Furdek, M., Bosiljevac, M., Skorin-Kapov, N. & Šipuš, Z. (2010a). Gain Competition in Optical Amplifiers: A Case Study. *Proceedings of International Convention on Information and Communication Technology, Electronics and Microelectronics* (MIPRO 2010), ISBN 978-1-4244-7763-0, Opatija, Croatia, May 2010

Furdek, M., Skorin-Kapov, N., Bosiljevac, M. & Šipuš, Z. (2010b). Analysis of Crosstalk in Optical Couplers and Associated Vulnerabilities. *Proceedings of International Convention on Information and Communication Technology, Electronics and Microelectronics* (MIPRO 2010), ISBN 978-1-4244-7763-0, Opatija, Croatia, May 2010

Furdek, M., Skorin-Kapov, N. & Grbac, M. (2010c). Attack-Aware Wavelength Assignment for Localization of In-band Crosstalk Attack Propagation. *IEEE/OSA Journal of Optical Communications and Networking*, Vol. 2, No. 11, (November 2010), pp. (1000-1009), ISSN 1943-0620

Ho, S.-T. & Chen, L.-K. (2009). Monitoring of Linearly Accumulated Optical Impairments in All-Optical Networks. *IEEE/OSA Journal of Optical Communications and Networking*, Vol. 1, No. 1, (June 2009), pp.(125-141), ISSN 1943-0620

Islam, M. N. (2003). Information Assurance and System Survivability in All-Optical Networks, Available from <www.eecs.umich.edu/OSL/Islam/SecureComm-WP.pdf>

Karásek, M. & Vallés, J.A. (1998). Analysis of Channel Addition/Removal Response in All-Optical Gain- Controlled Cascade of Erbium-Doped Fiber Amplifiers. *Journal of Lightwave Technologies*, Vol. 16, No. 10, (October 1998), pp. (1795-1803), ISSN 0733-8724.

Kilper, D.C., Bach, R., Blumenthal, D.J., Einstein, D., Landolsi, T., Ostar, L., Preiss, M. & Willner, A. E. (2004). Optical Performance Monitoring. *Journal of Lightwave Technologies*, Vol. 22, No. 1, (January 2004), pp. (294-304), ISSN 0733-8724.

Laude, J.-P. (2002). *DWDM Fundamentals, Components, and Applications*, Artech House, Inc., ISBN 1-58053-177-6, Norwood

Li, G., Yates, J., Wang, D. & Kalmanek, C. (2002). Control Plane Design for Reliable Optical Networks. *IEEE Communications Magazine*, Vol. 40, No. 2, (February 2002), pp. (90-96), ISSN 0136-6804

Liu, G. & Ji, C. (2007). Resilience of All-Optical Network Architectures under In-Band Crosstalk Attacks: A Probabilistic Graphical Model Approach. *IEEE Journal on Selected Areas in Communications*, Vol. 25, No. 4, (April 2007), pp. (2-17), ISSN 0733-8716

Mas, C., Tomkos, I. & Tonguz, O. (2005). Failure Location Algorithm for Transparent Optical Networks. *IEEE Journal on Selected Areas in Communications*, Vol. 23, No. 8, (August 2005), pp. (1508-1519), ISSN 0733-8716

Médard, M., Marquis, D., Barry, R.A. & Finn, S.G. (1997). Security Issues in All-Optical Networks. *IEEE Network*, Vol. 11, No. 3, (May/June 1997), pp. (42-48), ISSN 0890-8044

Médard, M., Marquis, D. & Chinn, S.R. (1998). Attack Detection Methods for All-Optical Networks, *Proceedings of Network and Distributed System Symposium (NDSS '98)*, ISBN 1-891562-01-0, San Diego, USA, March 1998.

Miller, S.K. (10 July 2007). Fiber Optic Security a Necessity, In: *SearchTelecom.com*, Available from <http://searchtelecom.techtarget.com/news/1263785/Fiber-optic-network-security-a-necessity>

Mukherjee, B. (2006.) *Optical WDM Networks*, Springer Science+Business Media, Inc., ISBN 978-0387-29055-3, New York.

Oyster Optics, Inc. (2002) Securing Fiber Optic Communications against Optical Tapping Methods, Available from <http://www.rootsecure.net/content/downloads/pdf/fiber_optic_taps.pdf>

Papadimitriou, G.I., Papazoglou, C. & Pomportsis, A.S. (2003). Optical Switching: Switch Fabrics, Techniques and Architectures. *Journal of Lightwave Technology*, Vol. 21, No. 2, (February 2003), pp. (384-405), ISSN 0733-8724

Peng, Y., Sun, Z., Du, S. & Long, K. (2011). Propagation of All-Optical Crosstalk Attack in Transparent Optical Networks. *Optical Engineering*, Vol. 50, No. 8, (August 2011), 085002, ISSN 0091-3286

Ramaswami, R. & Sivarajan, K.N. (2002). *Optical Networks: A Practical Perspective* (2nd edition), Morgan Kaufmann Publishers, ISBN 1-55860-655-6, San Francisco

Rejeb, R., Leeson, M.S & Green, R.J. (2006a). Multiple Attack Localization and Identification in All-Optical Networks. *Optical Switching and Networking*, Vol. 3, No. 1, (July 2006), pp. (41-49), ISSN 1573-4277

Rejeb, R., Leeson, M.S. & Green, R.J. (2006b). Fault and Attack Management in All-Optical Networks. *IEEE Communications Magazine*, Vol. 44, No. 11, (November 2006), pp. (79-86), ISSN 0163-6804

Rejeb, R., Leeson, M.S., Mas Machuca, C. & Tomkos, I. (2010). Control and Management Issues in All-Optical Networks. *Journal of Networks*, Vol. 5, No. 2, (February 2010), pp. (132-139), ISSN 1796-2056

Richardson, R. (2008). CSI Computer Crime & Security Survey, Available from: <http://gocsi.com/sites/default/files/uploads/CSIsurvey2008.pdf>

Rohit, A., Albores-Mejia, A., Calabretta, N., Leijtens, X., Robbins, D.J., Smit, M.K. & Williams, K. (2011). Fast Remotely Reconfigurable Wavelength Selective Switch, *Proceedings of Optical Fiber Communication Conference* (OFC 2011), Los Angeles, USA, ISBN 978-1-4577-0213-6

Saha, D., Rajagopalan, B. & Bernstein, G. (2003). The optical network control plane: state of the standards and deployment. *IEEE Communications Magazine*, Vol. 41, No. 8, (August 2003), pp. (S29-S34), ISSN 0163-6804

Shen, G. & Tucker, R.S. (2007.) Translucent Optical Networks: The Way Forward. *IEEE Topics in Optical Communications*, Vol. 45, No. 2, (February 2007), pp. (48-54), ISSN 0163-6804

Shuto, Y., Yanagi, S., Asakawa, S., Kobayashi, M. & Nagase, R. (2004). Fiber Fuse Phenomenon in Step-index Single-mode Optical Fibers. *IEEE Journal of Quantum Electronics*, Vol. 40, No. 8, (August 2004), pp. (1113-1121), ISSN 0018-9197

Skorin-Kapov, N., Chen, J. & Wosinska, L. (2010). A New Approach to Optical Networks Security: Attack-Aware Routing and Wavelength Assignment. *IEEE/ACM Transactions on Networking*, Vol. 18, No. 3, (June 2010), pp. (750-760), ISSN 1063-6692

Stanic, S. & Subramaniam, S. (2011). Fault Localization in All-Optical Networks with User and Supervisory Lightpaths. *Proceedings of IEEE International Conference on Communications (ICC 2011)*, ISBN 978-1-61284-232-5, Kyoto, Japan, June 2011

Tzanakaki, A., Zacharopoulos, I. & Tomkos, I. (2004). Broadband Building Blocks [Optical Networks]. *IEEE Circuits and Devices Magazine*, Vol. 20, No. 2, (March/April 2004), pp. (32-37), ISSN 8755-3996

Vaez, M.M. & Lea, C.-T. (2000). Strictly Nonblocking Directional-Coupler-Based Switching Networks Under Crosstalk Constraint, *IEEE Transactions on Networking*, Vol. 48, No. 2, (February 2000), pp. (316-323), ISSN 1036-6692

Way, I.W., Chen, D., Saifi, M.A., Andrejco, M.J., Yi-Yan, A., von Lehman, A.& Lin, C. (1991). High Gain Limiting Erbium-Doped Fiber Amplifier With Over 30 dB Dynamic Range, *IEEE Electronic Letters*, Vol. 27, No. 3, (January 1991), pp. (211-213), ISSN 0013-5194

Witcher, K. (2005). Fiber Optics and its Security Vulnerabilities. SANS Institute, Available from: <http://www.sans.org/reading_room/whitepapers/physcial/>

Wu, T. & Somani, A.K. (2005). Cross-Talk Attack Monitoring and Localization in All-Optical Networks, *IEEE/ACM Transactions on Networking*, Vol. 13, No. 6, (December 2005), pp.(1390-1401), ISSN 1036-6692

Zsigmond, S. (2011). External Report on Physical-Layer Attacks in Optical Networks. Technical report, project SAFE (http://www.fer.unizg.hr/tel/en/research/safe), supported by the Unity through Knowledge Fund (UKF), Ministry of Science, Education and Sports, Croatia, 2011

Part 3

Optical Communications Systems:
Multiplexing and Demultiplexing

7

Optical Demultiplexing Based on Four-Wave Mixing in Semiconductor Optical Amplifiers

Narottam Das[1,2] and Hitoshi Kawaguchi[3]
[1]Department of Electrical and Computer Engineering, Curtin University,
[2]School of Engineering, Edith Cowan University,
[3]Graduate School of Materials Science, Nara Institute of Science and Technology,
[1,2]Australia
[3]Japan

1. Introduction

Four-wave mixing (FWM) in semiconductor optical amplifiers (SOAs) has several important features, such as, high speed and high FWM conversion efficiency as well as optical demultiplexing (DEMUX) (Mecozzi et al., 1995; Mecozzi & Mørk, 1997; Das et al., 2000). The are several applications of FWM in SOAs for all-optical devices, such as, wavelength converters (Vahala et al., 1996), optical samplers (Inoue & Kawaguchi, 1998), optical phase conjugators (Kikuchi & Matsumura, 1998) and optical multiplexers/demultiplexers (Kawanishi et al., 1997; Kawanishi et al., 1994; Morioka et al., 1996; Uchiyama et al., 1998; Tomkos et al., 1999; Kirita et al., 1998; Buxens et al., 2000) have been demonstrated for optical communication systems. When a pulse of a time-multiplexed signal train (for example, a probe pulse) and a pump pulse are injected simultaneously into an SOA, gain and refractive index in the SOA are modulated and an FWM signal is generated by the modulations. Thus, we can obtain a demultiplexed signal as an FWM signal at the output of SOA. All-optical demultiplexing has been experimentally demonstrated up to 200 Gbit/s (Morioka et al., 1996). Tomkos et al., (Tomkos et al., 1999) suggested a number of ways to improve the performance of the dual-pump demultiplexer at 40 Gbit/s as follows; adjustment of the input wavelengths at the peak gain wavelength of the SOA under saturation conditions, the use of higher pump power at the input of the device, or/and the use of pulsed pumps with short pulsewidths. For the higher bit-rate, the overlap of the input to the FWM signal pulses may appear both in the time and spectral domain. The pattern effect may also appear in the FWM signal due to the slow components of the optical nonlinearities in SOAs (Saleh & Habbab, 1990). These effects degrade the usefulness of the FWM in SOAs as a practical DEMUX device in optical network/ communication systems. Therefore, it is very important to analysis the optical DEMUX characteristics based on FWM in SOAs for the ultrafast multi-bit input optical pulses.

The analyses of FWM in SOAs between short optical pulses have been widely reported (Shtaif & Eisenstein, 1995; Shtaif et al., 1995; Xie et al., 1999; Tang & Shore, 1998; Tang & Shore, 1999a; Tang & Shore, 1999b; Das et al., 2000). The FWM conversion efficiency (Shtaif et al., 1995; Xie et al., 1999; Tang & Shore, 1998; Tang & Shore, 1999a; Mørk & Mecozzi, 1997) the chirp of mixing pulses (Tang & Shore, 1999b; Das *et al.*, 2000), and the pump-probe time delay dependency of the FWM conversion efficiency (Shtaif & Eisenstein, 1995; Shtaif *et al.*, 1995; Mecozzi & Mørk, 1997; Das *et al.*, 2007; Das *et al.*, 2011) have been reported. On the contrary, however, there are only a few reports on analyses of FWM in SOAs used for demultiplexing time-division multiplexed data streams at ultra-high bit rates. Eiselt (Eiselt, 1995) reported the optimum control pulse energy and width with respect to the switching efficiency, channel crosstalk, and jitter tolerance. In those calculations, a very simple model of time-resolved gain saturation was used, which only took into account the gain recovery time. The FWM model was also very simple, in which the optical output power of the converted signal was proportional to the product of the squared pump output power and signal output power. Shtaif and Eisenstein (Shtaif & Eisenstein, 1996) calculated the error probabilities for time-domain DEMUX. Therefore, a detail and accurate analysis is required in order to clarify the performance of optical DEMUX based on FWM in SOAs for high-speed optical communication systems.

In this Chapter, we present detail numerical modeling/simulation results of FWM characteristics for the solitary probe pulse and optical DEMUX characteristics for multi-bit (multi-probe and/or pump) pulses in SOAs by using the finite-difference beam propagation method (FD-BPM) (Das *et al.*, 2000; Razaghi *et al.*, 2009). These simulations are based on the nonlinear propagation equation considering the group velocity dispersion, self-phase modulation (SPM), and two-photon absorption (TPA), with the dependencies on the carrier depletion (CD), carrier heating (CH), spectral-hole burning (SHB), and their dispersions, including the recovery times in SOAs (Hong *et al.*, 1996). For the simulation of solitary probe pulse, we obtain an optimum input pump pulsewidth from a viewpoint of ON/OFF ratios. For the simulation of optical DEMUX characteristics, we evaluate the ON/OFF ratios and the pattern effect of FWM signals for the multi-probe pulses. We have also simulated the optical DEMUX characteristics for the time-multiplexed signals by the repetitive pump pulses.

The FD-BPM is useful to obtain the propagation characteristics of single pulse or milti-pulses using the modified nonlinear Schrödinger equation (MNLSE) (Hong *et al.*, 1996 & Das *et al.*, 2000), simply by changing only the combination of input optical pulses. These are: (1) single pulse propagation (Das *et al.*, 2008), (2) FWM characteristics using two input pulses (Das *et al.*, 2000), (3) optical DENUX using several input pulses (Das *et al.*, 2001), (4) optical phase-conjugation using two input pulses with chirp (Das *et al.*, 2001) and (5) optimum time-delayed FWM characteristics between the two input pump and probe pulses (Das *et al.*, 2007).

2. Analytical model

In this section, we briefly discuss the important nonlinear effects in SOAs, mathematical formulation of modified nonlinear Schrödinger equation (MNLSE), finite-difference beam propagation method (FD-BPM) used in the simulation, and nonlinear propagation of solitary pulses in SOAs.

2.1 Important nonlinear effects in SOAs

There are several types of "nonlinear effects" in SOAs. Among them, the important four "nonlinear effects" are shown in Fig. 1. These are (i) spectral hole-burning (SHB), (ii) carrier heating (CH), (iii) carrier depletion (CD) and (iv) two-photon absorption (TPA).

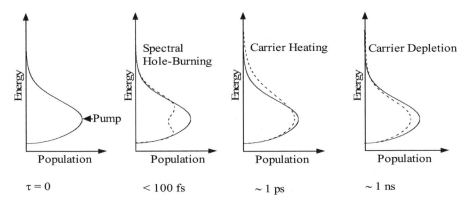

Fig. 1. Important nonlinear effects in SOAs are: (i) spectral hole-burning (SHB) with a life time of < 100 fs, (ii) carrier heating (CH) with a life time of ~ 1 ps, (iii) carrier depletion (CD) with a life time is ~ 1 ns and (iv) two-photon absorption (TPA).

Figure 1 shows the time-development of the population density in the conduction band after excitation (Das, 2000). The arrow (pump) shown in Fig. 1 is the excitation laser energy. Below the life-time of 100 fs, the SHB effect is dominant. SHB occurs when a narrow-band strong pump beam excites the SOA, which has an inhomogeneous broadening. SHB arises due to the finite value of intraband carrier-carrier scattering time (~ 50 – 100 fs), which sets the time scale on which a quasi-equilibrium Fermi distribution is established among the carriers in a band. After ~1 ps, the SHB effect is relaxed and the CH effect becomes dominant. The process tends to increase the temperature of the carriers beyond the lattice's temperature. The main causes of heating the carriers are (1) the stimulated emission, since it involves the removal of "cold" carriers close to the band edge and (2) the free-carrier absorption, which transfers carriers to high energies within the bands. The "hot"-carriers relax to the lattice temperature through the emission of optical phonons with a relaxation time of ~ 0.5 – 1 ps. The effect of CD remains for about 1 ns. The stimulated electron-hole recombination depletes the carriers, thus reducing the optical gain. The band-to-band relaxation also causes CD, with a relaxation time of ~ 0.2 – 1 ns. For ultrashort optical pumping, the two-photon absorption (TPA) effect also becomes important. An atom makes a transition from its ground state to the excited state by the simultaneous absorption of two laser photons. All these nonlinear effects (mechanisms) are taken into account in the simulation and the mathematical formulation of modified nonlinear Schrödinger equation (MNLSE).

2.2 Mathematical formulation of modified nonlinear schrödinger equation (MNLSE)

In this subsection, we will briefly explain the theoretical analysis of short optical pulses propagation in SOAs. We start from Maxwell's equations (Agrawal, 1989; Yariv, 1991; Sauter, 1996) and reach the propagation equation of short optical pulses in SOAs, which are governed by the wave equation (Agrawal & Olsson, 1989) in the frequency domain:

$$\nabla^2 \overline{E}(x,y,z,\omega) + \frac{\varepsilon_r}{c^2} \omega^2 \overline{E}(x,y,z,\omega) = 0 \tag{1}$$

where, $\overline{E}(x,y,z,\omega)$ is the electromagnetic field of the pulse in the frequency domain, c is the velocity of light in vacuum and ε_r is the nonlinear dielectric constant which is dependent on the electric field in a complex form. By slowly varying the envelope approximation and integrating the transverse dimensions we arrive at the pulse propagation equation in SOAs (Agrawal & Olsson, 1989; Dienes et al., 1996).

$$\frac{\partial V(\omega,z)}{\partial z} = -i\left\{\frac{\omega}{c}[1+\chi_m(\omega)+\Gamma\tilde{\chi}(\omega,N)]^{1/2}-\beta_0\right\}V(\omega,z) \tag{2}$$

where, $V(\omega,z)$ is the Fourier-transform of $V(t,z)$ representing pulse envelope, $\chi_m(\omega)$ is the background (mode and material) susceptibility, $\tilde{\chi}(\omega)$ is the complex susceptibility which represents the contribution of the active medium, N is the effective population density, β_0 is the propagation constant. The quantity Γ represents the overlap/ confinement factor of the transverse field distribution of the signal with the active region as defined in (Agrawal & Olsson, 1989).

Using mathematical manipulations (Sauter, 1996; Dienes et al., 1996), including the real part of the instantaneous nonlinear Kerr effect as a single nonlinear index n_2 and by adding the two-photon absorption (TPA) term we obtain the MNLSE for the phenomenological model of semiconductor laser and amplifiers (Hong et al., 1996). The following MNLSE (Hong et al., 1996; Das et al., 2000) is used for the simulation of FWM characteristics with solitary probe pulse and optical DEMUX characteristics with multi-probe or pump in SOAs:

$$\left[\frac{\partial}{\partial z}-\frac{i}{2}\beta_2\frac{\partial^2}{\partial\tau^2}+\frac{\gamma}{2}+\left(\frac{\gamma_{2p}}{2}+ib_2\right)|V(\tau,z)|^2\right]V(\tau,z)$$

$$=\left\{\frac{1}{2}g_N(\tau)\left[\frac{1}{f(\tau)}+i\alpha_N\right]+\frac{1}{2}\Delta g_T(\tau)(1+i\alpha_T)-i\frac{1}{2}\frac{\partial g(\tau,\omega)}{\partial\omega}\bigg|_{\omega_0}\frac{\partial}{\partial\tau}-\frac{1}{4}\frac{\partial^2 g(\tau,\omega)}{\partial\omega^2}\bigg|_{\omega_0}\frac{\partial^2}{\partial\tau^2}\right\}V(\tau,z) \tag{3}$$

We introduce the frame of local time τ (=t - z/v_g), which propagates with a group velocity v_g at the center frequency of an optical pulse. A slowly varying envelope approximation is used in (3), where the temporal variation of the complex envelope function is very slow compared with the cycle of the optical field. In (3), $V(\tau,z)$ is the time domain complex envelope function of an optical pulse, $|V(\tau,z)|^2$ corresponding to the optical power, and β_2 is the GVD. γ is the linear loss, γ_{2p} is the two-photon absorption coefficient, b_2 (= $\omega_0 n_2/cA$) is the instantaneous self-phase modulation term due to the instantaneous nonlinear Kerr effect n_2, ω_0 (= $2\pi f_0$) is the center angular frequency of the pulse, c is the velocity of light in vacuum, A (= wd/Γ) is the effective area (d and w are the thickness and width of the active region, respectively and Γ is the confinement factor) of the active region.

The saturation of the gain due to the CD is given by (Hong et al., 1996)

$$g_N(\tau) = g_0 \exp\left(-\frac{1}{W_s}\int_{-\infty}^{\tau} e^{-s/\tau_s}|V(s)|^2\,ds\right) \tag{4}$$

where, $g_N(\tau)$ is the saturated gain due to CD, g_0 is the linear gain, W_s is the saturation energy, τ_s is the carrier lifetime.

The SHB function $f(\tau)$ is given by (Hong et al., 1996)

$$f(\tau) = 1 + \frac{1}{\tau_{shb}P_{shb}} \int_{-\infty}^{+\infty} u(s)e^{-s/\tau_{shb}} |V(\tau-s)|^2 ds \qquad (5)$$

where, $f(\tau)$ is the SHB function, P_{shb} is the SHB saturation power, τ_{shb} is the SHB relaxation time, and α_N and α_T are the linewidth enhancement factor associated with the gain changes due to the CD and CH.

The resulting gain change due to the CH and TPA is given by (Hong et al., 1996)

$$\Delta g_T(\tau) = -h_1 \int_{-\infty}^{+\infty} u(s)e^{-s/\tau_{ch}}(1 - e^{-s/\tau_{shb}})|V(\tau-s)|^2 ds$$
$$- h_2 \int_{-\infty}^{+\infty} u(s)e^{-s/\tau_{ch}}(1 - e^{-s/\tau_{shb}})|V(\tau-s)|^4 ds \qquad (6)$$

where, $\Delta g_T(\tau)$ is the resulting gain change due to the CH and TPA, $u(s)$ is the unit step function, τ_{ch} is the CH relaxation time, h_1 is the contribution of stimulated emission and free-carrier absorption to the CH gain reduction and h_2 is the contribution of two-photon absorption.

The dynamically varying slope and curvature of the gain plays a shaping role for pulses in the sub-picosecond range. The first and second order differential net (saturated) gain terms are (Hong et al., 1996),

$$\left.\frac{\partial g(\tau,\omega)}{\partial \omega}\right|_{\omega_0} = A_1 + B_1[g_0 - g(\tau,\omega_0)] \qquad (7)$$

$$\left.\frac{\partial^2 g(\tau,\omega)}{\partial \omega^2}\right|_{\omega_0} = A_2 + B_2[g_0 - g(\tau,\omega_0)] \qquad (8)$$

$$g(\tau,\omega_0) = g_N(\tau,\omega_0) / f(\tau) + \Delta g_T(\tau,\omega_0) \qquad (9)$$

where, A_1 and A_2 are the slope and curvature of the linear gain at ω_0, respectively, while B_1 and B_2 are constants describing changes in A_1 and A_2 with saturation, as given in (7) and (8).

The gain spectrum of an SOA is approximated by the following second-order Taylor expansion in $\Delta\omega$:

$$g(\tau,\omega) = g(\tau,\omega_0) + \Delta\omega \left.\frac{\partial g(\tau,\omega)}{\partial \omega}\right|_{\omega_0} + \frac{(\Delta\omega)^2}{2}\left.\frac{\partial^2 g(\tau,\omega)}{\partial \omega^2}\right|_{\omega_0} \qquad (10)$$

The coefficients $\left.\dfrac{\partial g(\tau,\omega)}{\partial \omega}\right|_{\omega_0}$ and $\left.\dfrac{\partial^2 g(\tau,\omega)}{\partial \omega^2}\right|_{\omega_0}$ are related to A_1, B_1, A_2 and B_2 by (7) and (8).

Here we assumed the same values of A_1, B_1, A_2 and B_2 as in (Hong et al., 1996) for an AlGaAs/GaAs bulk SOA.

The time derivative terms in (3) have been replaced by the central-difference approximation in order to simulate this equation by the FD-BPM (Das et al., 2000). In simulation, the parameter of bulk SOAs (AlGaAs/GaAs, double heterostructure) with a wavelength of 0.86 μm (Hong et al., 1996) is used and the SOA length is 350 μm. The input pulse shape is sech² and is Fourier transform-limited.

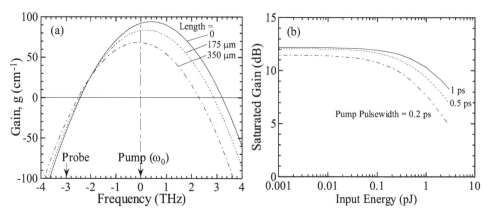

Fig. 2. (a) The gain spectra given by the second-order Taylor expansion about the center frequency of the pump pulse ω_0. The solid line shows the unsaturated gain spectrum (length: 0 μm), the dotted and the dashed-dotted lines are a saturated gain spectrum at 175 μm and 350 μm, respectively. Here, the input pump pulse pulsewidth is 1 ps and pulse energy is 1 pJ. (b) Saturated gain versus the input pump pulse energy characteristics of the SOA. The saturation energy decreases with decreasing the input pump pulsewidth. The SOA length is 350 μm. The input pulsewidths are 0.2 ps, 0.5 ps, and 1 ps respectively, and a pulse energy of 1 pJ.

The gain spectra of SOAs are very important for obtaining the propagation and wave mixing (FWM and optical DEMUX between the input pump and probe pulses) characteristics of short optical pulses. Figure 2(a) shows the gain spectra given by a second-order Taylor expansion about the pump pulse center frequency ω_0 with derivatives of $g(\tau, \omega)$ by (7) and (8) (Das et al., 2000). In Fig. 2(a), the solid line represents an unsaturated gain spectrum (length: 0 μm), the dotted line represents a saturated gain spectrum at the center position of the SOA (length: 175 μm), and the dashed–dotted line represents a saturated gain spectrum at the output end of the SOA (length: 350 μm), when the pump pulsewidth is 1 ps and input energy is 1 pJ. These gain spectra were calculated using (1), because, the waveforms of optical pulses depend on the propagation distance (i.e., the SOA length). The spectra of these pulses were obtained by Fourier transformation. The "local" gains at the center frequency at z = 0, 175, and 350 μm were obtained from the changes in the pulse intensities at the center frequency at around those positions (Das et al., 2001). The gain at the center frequency in the gain spectrum was approximated by the second-order Taylor expression series. As the pulse propagates in the SOA, the pulse intensity increases due to the gain of the SOA. The increase in pulse intensity reduces the gain, and the center frequency of the gain shifts to lower frequencies. The pump frequency is set to near the gain peak, and linear gain g_0 is 92 cm at ω_0. The probe frequency is set -3 THz from ω_0 for the calculations of FWM characteristics as described below, and the linear gain g_0 is -42 cm at this frequency. Although the probe frequency lies outside the gain bandwidth, we selected a detuning of 3 THz in this simulation because the FWM signal must be spectrally separated from the output of the SOA. As will be shown later, even for this large degree of detuning, the FWM signal pulse and the pump pulse spectrally overlap when the pulsewidths become short (<0.5 ps) (Das *et al.*, 2001). The gain bandwidth is about the same as the measured value for an AlGaAs/GaAs bulk SOA (Seki *et al.*, 1981). If an InGaAsP/InP bulk SOA is used we can expect much wider gain bandwidth (Leuthold *et al.*, 2000). With a decrease in the carrier density, the gain decreases and the peak position is shifted to a lower frequency because of the band-filling effect. Figure 2(b) shows the saturated gain versus input pump pulse energy characteristics of the SOA. When the input pump pulsewidth decreases then the small signal gain decreases due to the spectral limit of the gain bandwidth. For the case, when the input pump pulsewidth is short (very narrow, such as 200 fs or lower), the gain saturates at small input pulse energy (Das *et al.*, 2000). This is due to the CH and SHB with the fast response.

Initially, the MNLSE was used by (Hong *et al.*, 1996) for the analysis of "solitary pulse" propagation in an SOA. We used the same MNLSE for the simulation of FWM and optical DEMUX characteristics in SOA using the FD-BPM. Here, we have introduced a complex envelope function $V(\tau, 0)$ at the input side of the SOA for taking into account the two (pump and probe) or more (multi-pump or probe) pulses.

2.3 Finite-difference beam propagation method (FD-BPM)

To solve a boundary value problem using the finite-differences method, every derivative term appearing in the equation, as well as in the boundary conditions, is replaced by the central differences approximation. Central differences are usually preferred because they lead to an excellent accuracy (Conte & Boor, 1980). In the modeling, we used the finite-differences (central differences) to solve the MNLSE for this analysis.

Usually, the fast Fourier transformation beam propagation method (FFT-BPM) (Okamoto, 1992; Brigham, 1988) is used for the analysis of the optical pulse propagation in optical fibers by the successive iterations of the Fourier transformation and the inverse Fourier transformation. In the FFT-BPM, the linear propagation term (GVD term) and phase compensation terms (other than GVD, 1st and 2nd order gain spectrum terms) are separated in the nonlinear Schrödinger equation for the individual consideration of the time and frequency domain for the optical pulse propagation. However, in our model, equation (3) includes the dynamic gain change terms, i.e., the 1st and 2nd order gain spectrum terms which are the last two terms of the right-side in equation (3). Therefore, it is not possible to separate equation (3) into the linear propagation term and phase compensation term and it is quite difficult to calculate equation (3) using the FFT-BPM. For this reason, we used the FD-BPM (Chung & Dagli, 1990; Conte & Boor, 1980; Das *et al.*, 2000; Razaghi *et al.*, 2009). If we replace the time derivative terms of equation (3) by the below central-difference approximation, equation (11), and integrate equation (3) with the small propagation step Δz, we obtain the tridiagonal simultaneous matrix equation (12)

$$\frac{\partial}{\partial \tau} V_k = \frac{V_{k+1} - V_{k-1}}{2\Delta\tau}, \quad \frac{\partial^2}{\partial\tau^2} V_k = \frac{V_{k+1} - 2V_k + V_{k-1}}{\Delta\tau^2} \tag{11}$$

where, $V_k = V(\tau_k)$, $V_{k+1} = V(\tau_k + \Delta\tau)$, and $V_{k-1} = V(\tau_k - \Delta\tau)$

$$-a_k(z + \Delta z) V_{k-1}(z + \Delta z) + \{1 - b_k(z + \Delta z)\} V_k(z + \Delta z) - c_k(z + \Delta z) V_{k+1}(z + \Delta z)$$
$$= a_k(z) V_{k-1}(z) + \{1 + b_k(z)\} V_k(z) + c_k(z) V_{k+1}(z) \tag{12}$$

where, $k = 1, 2, 3, \ldots\ldots\ldots, n$ and

$$a_k(z) = \frac{\Delta z}{2}\left[\frac{i\beta_2}{2\Delta\tau^2} + i\frac{1}{4\Delta\tau}\frac{\partial g(\tau, \omega, z)}{\partial\omega}\Big|_{\omega_0, \tau_k} - \frac{1}{4\Delta\tau^2}\frac{\partial^2 g(\tau, \omega, z)}{\partial\omega^2}\Big|_{\omega_0, \tau_k} \right] \tag{13}$$

$$b_k(z) = -\frac{\Delta z}{2}\left[\frac{i\beta_2}{\Delta\tau^2} + \frac{\gamma}{2} + \left(\frac{\gamma_{2p}}{2} + ib_2\right)|V_k(z)|^2 - \frac{1}{2}g_N(\tau_k, \omega_0, z)(1 + i\alpha_N) \right.$$
$$\left. -\frac{1}{2}\Delta g_T(\tau_k, \omega_0, z)(1 + i\alpha_T) - \frac{1}{2\Delta\tau^2}\frac{\partial^2 g(\tau, \omega, z)}{\partial\omega^2}\Big|_{\omega_0, \tau_k} \right] \tag{14}$$

$$c_k(z) = \frac{\Delta z}{2}\left[\frac{i\beta_2}{2\Delta\tau^2} - i\frac{1}{4\Delta\tau}\frac{\partial g(\tau, \omega, z)}{\partial\omega}\Big|_{\omega_0, \tau_k} - \frac{1}{4\Delta\tau^2}\frac{\partial^2 g(\tau, \omega, z)}{\partial\omega^2}\Big|_{\omega_0, \tau_k} \right] \tag{15}$$

where, $\Delta\tau$ is the sampling time and n is the number of sampling. If we know $V_k(z)$, ($k = 1, 2, 3, \ldots\ldots, n$) at the position z, we can calculate $V_k(z + \Delta z)$ at the position of $z + \Delta z$ which is the propagation of a step Δz from position z, by using equation (12). It is not possible to directly calculate equation (12) because it is necessary to calculate the left-side terms $a_k(z + \Delta z)$, $b_k(z + \Delta z)$, and $c_k(z + \Delta z)$ of equation (12) from the unknown $V_k(z + \Delta z)$. Therefore, we initially defined $a_k(z + \Delta z) \equiv a_k(z)$, $b_k(z + \Delta z) \equiv b_k(z)$, and $c_k(z + \Delta z) \equiv c_k(z)$ and obtained $V_k^{(0)}(z + \Delta z)$, as the zeroth order approximation of $V_k(z + \Delta z)$ by using equation (12). We then substituted $V_k^{(0)}(z + \Delta z)$ in equation (12) and obtained $V_k^{(1)}(z + \Delta z)$ as the first order approximation of $V_k(z + \Delta z)$ and finally obtained the accurate simulation results by the iteration as used in (Brigham, 1988; Chung & Dagli, 1990; Das et al., 2000; Razaghi et al., 2009).

Figure 3 shows a simple schematic diagram of the FD-BPM in time domain. Here, $\tau (= t - z / v_g)$ is the local time, which propagates with the group velocity v_g at the center frequency of an optical pulse and $\Delta\tau$ is the sampling time. z is the propagation direction and Δz is the propagation step. With this procedure, we used up to 3-rd time iteration for more accuracy of the simulations.

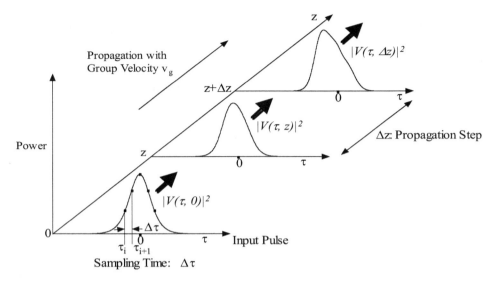

Fig. 3. A simple schematic diagram of FD-BPM in the time domain, where, $\tau \, (= t - z / v_g)$ is the local time, which propagates with the group velocity v_g at the center frequency of an optical pulse and $\Delta\tau$ is the sampling time, and z is the propagation direction and Δz is the propagation step.

The FD-BPM (Conte & Boor, 1980; Chung & Dagli, 1990; Das et al., 2000; Razaghi et al., 2009a & 2009b) is used for the simulation of several important characteristics, namely, (1) single pulse propagation in SOAs (Das et al., 2008; Razaghi et al., 2009a & 2009b), (2) two input pulses propagating in SOAs (Das et al., 2000; Connelly et al., 2008), (3) Optical DEMUX characteristics of multi-probe or pump input pulses based on FWM in SOAs (Das et al., 2001), (4) Optical phase-conjugation characteristics of picosecond FWM signal in SOAs (Das et al., 2001), and (5) FWM conversion efficiency with optimum time-delays between the input pump and probe pulses (Das et al., 2007).

2.4 Nonlinear optical pulse propagation model in SOAs

Nonlinear optical pulse propagation in SOAs has drawn considerable attention due to its potential applications in optical communication systems, such as a wavelength converter based on FWM and switching. The advantages of using SOAs include the amplification of small (weak) optical pulses and the realization of high efficient FWM characteristics.

For the analysis of optical pulse propagation in SOAs using the FD-BPM in conjunction with the MNLSE, where several parameters are taken into account, namely, the group velocity dispersion, self-phase modulation (SPM), and TPA, as well as the dependencies on the CD, CH, SHB and their dispersions, including the recovery times in an SOA (Hong et al., 1996). We also considered the gain spectrum (as shown in Fig. 2). The gain in an SOA was dynamically changed depending on values used for the carrier density and carrier temperature in the propagation equation (i.e., MNLSE).

Initially, (Hong et al., 1996) used the MNLSE for the simulation of optical pulse propagation in an SOA by FFT-BPM (Okamoto, 1992; Brigham, 1988) but the dynamic gain terms were changing with time. The FD-BPM is capable to simulate the optical pulse propagation taking into consideration the dynamic gain terms in SOAs (Das et al., 2000 & 2007; Razaghi et al., 2009a & 2009b; Aghajanpour et al., 2009). We used the MNLSE for nonlinear optical pulse propagation in SOAs by the FD-BPM (Chung & Dagli, 1990; Conte & Boor, 1980). We used the FD-BPM for the simulation of optical DEMUX characteristics in SOAs with the multi-input pump and probe pulses.

Fig. 4. A simple schematic diagram for the simulation of nonlinear single pulse propagation in SOA. Here, $|V(\tau,0)|^2$ is the input (z = 0) pulse intensity and $|V(\tau,z)|^2$ is the output pulse intensity (after propagating a distance z) of SOA.

Figure 4 illustrates a simple model for the simulation of nonlinear optical pulse propagation in an SOA. An optical pulse is injected into the input side of the SOA (z = 0). Here, τ is the local time, $|V(\tau,0)|^2$ is the intensity (power) of input pulse (z = 0) and $|V(\tau,z)|^2$ is the intensity (power) of the output pulse after propagating a distance z at the output side of SOA. We used this model to simulate FWM (with single probe) and DEMUX (with multi-bit pump or probe pulses) characteristics in SOAs.

3. FWM characteristics in SOAs with the solitary probe pulse

In this section, we will discuss the FWM characteristics with the solitary probe pulse in SOAs. When two optical pulses with different central frequencies f_p (pump) and f_q (probe) are injected simultaneously into the SOA, an FWM signal is generated at the output of the SOA at a frequency of $2f_p - f_q$ (as shown in Fig. 5). For the analysis (simulation) of FWM characteristics, the total input pump and probe pulse, $V_{in}(\tau)$, is given by the following equation

$$V_{in}(\tau) = V_p(\tau) + V_q(\tau)\exp(-i2\pi\Delta f\tau) \tag{16}$$

where, $V_p(\tau)$ and $V_q(\tau)$ are the complex envelope functions of the input pump and probe pulses respectively, $\tau\,(=t-z/v_g)$ is the local time that propagates with group velocity v_g at the center frequency of an optical pulse, Δf is the detuning frequency between the input pump and probe pulses and expressed as $\Delta f = f_p - f_q$. Using the complex envelope function of (16), we solved the MNLSE and obtained the combined spectrum of the amplified pump, probe and the generated FWM signal at the output of SOA.

Name of the Parameters	Symbols	Values	Units
Length of SOA	L	350	μm
Effective area	A	5	μm²
Center frequency of the pulse	f_0	349	THz
Linear gain	g_0	92	cm⁻¹
Group velocity dispersion	β_2	0.05	ps² cm⁻¹
Saturation energy	W_s	80	pJ
Linewidth enhancement factor due to the CD	α_N	3.1	
Linewidth enhancement factor due to the CH	α_T	2.0	
The contribution of stimulated emission and FCA to the CH gain reduction	h_1	0.13	cm⁻¹pJ⁻¹
The contribution of TPA	h_2	126	fs cm⁻¹pJ⁻²
Carrier lifetime	τ_s	200	ps
CH relaxation time	τ_{ch}	700	fs
SHB relaxation time	τ_{shb}	60	fs
SHB saturation power	P_{shb}	28.3	W
Linear loss	γ	11.5	cm⁻¹
Instantaneous nonlinear Kerr effect	n_2	-0.70	cm² TW⁻¹
TPA coefficient	γ_{2p}	1.1	cm⁻¹ W⁻¹
Parameters describing second-order Taylor expansion of the dynamically gain spectrum	A_1	0.15	fs μm⁻¹
	B_1	-80	fs
	A_2	-60	fs² μm⁻¹
	B_2	0	fs²

Table 1. Simulation parameters of a bulk SOA (AlGaAs/GaAs, double heterostructure) (Hong et al., 1996; Das et al., 2000).

For the simulations, we used the parameters of a bulk SOA (AlGaAs/GaAs, double heterostructure) at a wavelength of 0.86 μm. The parameters are listed in Table 1 (Hong et al., 1996). The length of the SOA was assumed to be 350 μm. All the results were obtained for a propagation step Δz of 5 μm. We confirmed that for any step size less than 5 μm the simulation results were almost identical (i.e., independent of the step size).

For the simulation of optical DEMUX characteristics in SOAs, we have started with the simulation of FWM characteristics for solitary probe pulse's. Figure 5 shows a simple schematic diagram illustrating the simulation of the FWM characteristics in an SOA between short optical pulses. In SOAs, the FWM signal is generated by mixing between the input pump and probe pulses, whose frequency appears at the symmetry position of the probe pulse with respect to the pump. We have selected the detuning frequency between the input pump and probe pulses to +3 THz. The generated FWM signal is filtered using an optical narrow bandpass filter from the optical output spectrum containing the pump and probe signal. Here, the pass-band of the filter is set to be from +2 THz to +4 THz, i.e., a bandwidth of 2 THz is used. The shape of the pass-band was assumed to be rectangular. The solid line represents for a short pump pulsewidths and the dotted line represents for a wider pump pulsewidths. For a wider pulsewidth, the pump peak intensity decreases, spectral peak intensity increases, and the FWM signal peak intensity decreases as shown in the figure when the input pulse energy is kept constant.

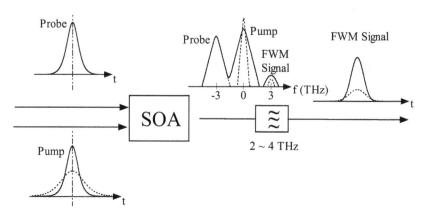

Fig. 5. A simple schematic diagram for the simulation of FWM characteristics for solitary probe pulse for the optimization of the input pump pulsewidth. Here, the input pump pulsewidth is varied.

Figure 6(a) shows the calculated output spectra of the SOA. The solid and dashed curves show the output spectra with pump and probe pulses and with only a pump pulse, respectively. The pump pulsewidths are 1 ps, 0.5 ps, and 0.2 ps and the probe pulsewidth is 1 ps. The input pump and probe pulse energies are 1 pJ and 10 fJ, respectively. For a pump pulsewidth of 1 ps, the spectral peaks of the pump, probe, and FWM signals are clearly separated. The FWM signal can be obtained by the spectral filtering whose bandwidth is from +2 THz to +4 THz, which is shown in the figure by the arrow. With the decrease in the pump pulsewidth, the pump spectral width is broadened and it becomes difficult to extract the FWM signal using the optical filter due to the spectral overlap. For the shorter pump pulsewidth less than 0.5 ps, the FWM signals are not clearly observed.

Figure 6(b) shows the temporal waveforms of the output signals after filtering. The solid and dashed curves show the waveforms with probe and without probe pulses, respectively. The contrast between the output power with probe and without probe pulses for a pump pulsewidth of 1 ps is larger than that for the shorter pump pulsewidths. This is due to the strong overlap between pump pulse of 0.5 ps and 0.2 ps and the FWM signal in the frequency domain. For an input pump pulsewidth of 1 ps, a FWM signal pulsewidth of 0.73 ps is narrower than the input pump pulsewidth. This is due to the fact that the FWM signal intensity is proportional to $I_p^2 I_q$ i.e., $I_{FWM} \propto I_p^2 I_q$ as reported (Das et $al.$, 2000). Here, I_{FWM} is the FWM signal intensity, I_p is the pump pulse intensity, and I_q is the probe pulse intensity. For input pump pulsewidths of 0.5 ps and 0.2 ps, the optical bandpass filter broadens the FWM signal pulsewidth due to the limitation in the frequency domain. Then, FWM signal pulsewidths of 0.57 ps and 0.55 ps become broader than the input pump pulsewidths. By this filtering, the energies of the transform-limited sech² pulses with pulsewidths of 1 ps, 0.73 ps, 0.5 ps, and 0.2 ps are reduced by 0.002%, 0.05%, 0.7%, and 19%, respectively. The peak powers are decreased by 0.86%, 4%, 14%, and 63%, respectively. The pulsewidth of 0.73 ps corresponds the that of the FWM pulse among 1 ps pump and 1 ps probe pulses (Fig. 6(b)). Therefore, the waveform distortion by this filtering is negligibly small for the FWM pulses among 1 ps pump and 1 ps probe pulses.

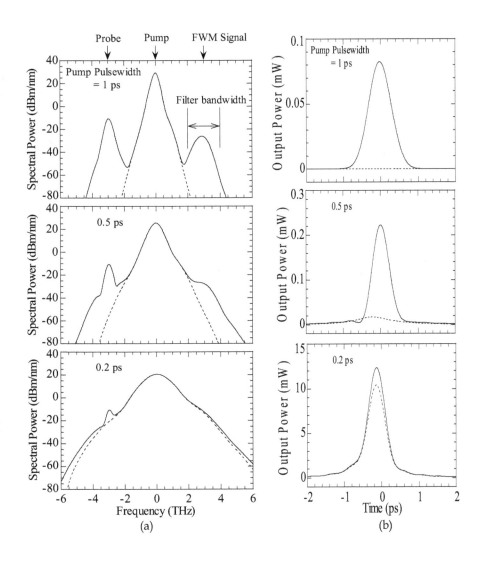

Fig. 6. (a) Output spectra of the SOA before filtering. The solid and dashed curves show the output spectra with pump and probe pulses and with only a pump pulse, respectively. (b) Output pulse waveforms after filtering from +2 ~ +4 THz. The solid and dashed curves show the output pulse waveforms with pump and probe pulses and with only a pump pulse, respectively. Here, the input pump and probe pulse energies are 1 pJ and 10 fJ, respectively. The input probe pulsewidth is 1 ps.

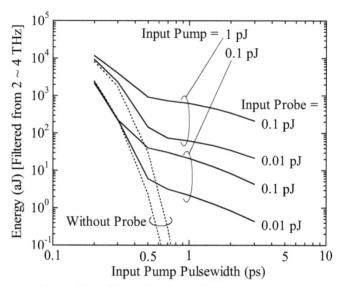

Fig. 7. The energy, which is filtered from the +2 to +4 THz component of the output spectrum, against the input pump pulsewidth. The input probe pulse energies are 0.01 pJ, and 0.1 pJ, and pulsewidth is 1 ps. The input pump energies are 0.1 pJ and 1 pJ. The solid lines represent the FWM signal energy with the input probe pulse and the dashed lines represent the FWM signal energy without input probe pulse.

Figure 7 shows the energy, which is filtered from the +2 to +4 THz component of the output spectrum (as shown in Fig. 6(a)) versus the input pump pulsewidth characteristics. The solid and dashed curves show the output pulse energy with and without probe pulses, respectively. Here, the input probe pulse energies were set to be 0.01 pJ, and 0.1pJ with a pulsewidth of 1 ps and the input pump pulse energies were set to be 0.1 pJ and 1 pJ. With the decrease of pump pulsewidth, the output energy increases, while the differences between the output energy with probe pulses and without probe pulses decrease because of the overlap of the pump and the FWM signal in the spectral domain. Therefore, the regular DEMUX operation is not obtained for the pump pulsewidth of less than 0.5 ps.

Figure 8(a) shows the ON-OFF ratio and the FWM conversion efficiency characteristics. Here, the "ON-OFF ratio" is defined as the ratio of the output energy having a spectral component of +2 ~ +4 THz with the probe pulse to without the probe pulse. Therefore, in the ideal case, the output energy with the probe pulse corresponds to the FWM signal energy and the output energy without the probe pulse becomes zero. The larger ON-OFF ratio is preferable in the DEMUX operations. We have assumed that the ON-OFF ratio > 20 dB is acceptable for a practical DEMUX operation. To obtain the enough ON-OFF ratio, the pump pulsewidth should be wider than 0.8 ps for input pump pulse energies of 0.1 pJ and 1 pJ. Fig. 8(b) shows the FWM conversion efficiency increases with the increase of input pump pulse energy. The FWM signal intensity is proportional to the square of the input pump intensity (Das et al., 2000). Therefore, the FWM conversion efficiency increases about 20 dB for the increase in a pump pulse energy of 10 dB in the region where the enough ON-OFF ratio is obtained. For the narrower pump pulsewidth, the nominal FWM conversion efficiency increases. However, the pump pulses narrower than 0.8 ps are not suitable for a practical DEMUX operation, which is due to the low ON-OFF ratio. To improve the ON-OFF ratio, one possible method is to increase the detuning for decreasing the spectral overlap between the pump and the FWM signal. However, in our simulation using the parameters for typical SOAs working in a 0.86 μm region, the increasing in the detuning is not practical because of the limited gain bandwidth of the SOAs. It is necessary to use the SOA with wider gain bandwidth for FWM among shorter pulses. One good candidate for obtaining wider gain bandwidth is to use the SOAs with staggered thickness multiple quantum wells (Mikami *et al.*,1991; Gingrich *et al.*, 1997).

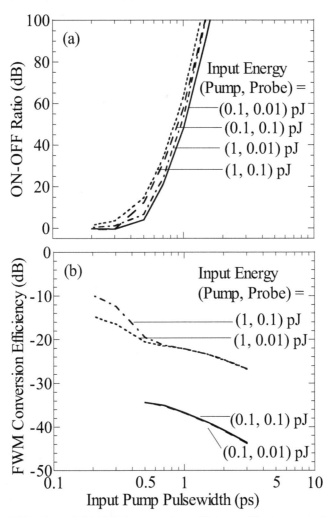

Fig. 8. (a) ON-OFF ratio and (b) FWM conversion efficiency. The input probe pulse energies are 0.01 pJ and 0.1 pJ, and pulsewidth is 1 ps. The input pump pulse energies are 0.1 pJ and 1 pJ.

4. Optical DEMUX characteristics in SOAs with multi-bit probe or pump pulses

In this section, we will discuss the optical DEMUX characteristics in SOAs with multi-bit probe or multi-bit pump pulses and vice versa. The FWM signal generates only when the pump and probe pulses are injected simultaneously into SOAs. Therefore, all-optical demultiplexed signals can be extracted as the FWM signals from a time-multiplexed signal train as described in the Introduction. Here, the pump and probe pulses act as gating and gated pulses, respectively. For a solitary probe pulse, the overlap between the pump pulse and the FWM signal in the frequency domain decreases the ON-OFF ratio as described in the previous section. The overlap in the frequency domain increases with the decrease in the pump pulsewidth. Therefore, the ON-OFF ratio increases with the increase of pump pulsewidth. On the otherhand, for the multi-bit probe pulses, the overlap among the pulses in the time domain also decreases the ON-OFF ratio. This overlap mainly comes from the neighboring pulses in the time domain. To investigate the influence of the neighboring pulses, we simulate the optical DEMUX characteristics for a three-bit-stream of 250 Gbit/s in this section. We also evaluate the pattern effect on the DEMUX (based on FWM) signals caused by the probe pulses.

Figure 9 shows the schematic diagram for the simulation of the ON-OFF ratio of the all-optical DEMUX. The probe pulses are a three-bit-stream of 250 Gbit/s. The peak position of the pump pulse is adjusted to that of the center pulse of the three probe pulses. Here, the ON-OFF ratio is defined as the ratio of the FWM signal energy obtained with the central input probe pulse of the three-bit-stream to the one obtained without the central input probe pulse. For the wider input pump pulse as indicated by the dashed lines, the FWM signal decreases and the ON-OFF ratio decreases, i.e., the crosstalk from the neighboring pulses increases.

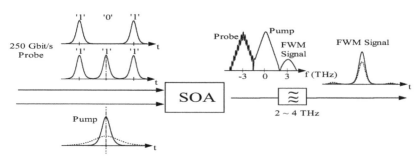

Fig. 9. A simple schematic diagram for the simulation of ON-OFF ratio of all-optical DEMUX. The input probe pulse repetition rate is 250 Gbit/s and pulsewidth is 1 ps. In the input probe pulse stream, '0' represents the signal is OFF and '1' represents the signal is ON.

Figure 10 shows the calculated ON-OFF ratio versus the input pump pulsewidth characteristics for the three-bit-stream. The input probe pulse energies are 0.1 pJ and 0.01 pJ, and pulsewidth is 1 ps. The input pump energies are 0.01 pJ, 0.1 pJ, and 1 pJ. For the wider input pump pulsewidth, the ON-OFF ratio decreases due to the overlap in the time domain among the pump and the neighboring of probe pulses. For an input pump pulsewidth of 3 ps, the ON-OFF ratio becomes about 20 dB. This relatively large allowance in the pump pulsewidth is due to the fact that the pulsewidth of the probe pulses (1 ps) is set to be short compared with a bit interval of 4 ps (250 Gbit/s). On the other hand, with the decreasing in the pump pulsewidth, the ON-OFF ratio severely decreases due to the overlap in the frequency domain between the pump and the FWM signal pulse as explained in Fig. 8. This small allowance is attributed to the fact that the pump pulse energy is much stronger than that of the FWM signal. These results have an interesting information; the overlap in the frequency domain is more important than the overlap in the time domain for the design of the ultrafast all-optical DEMUX. As a result of the simulation, the optimum input pump pulsewidth range is 1 ps ~ 3 ps for an input probe pulsewidth of 1 ps.

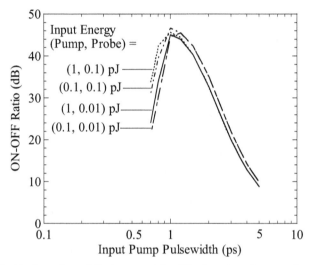

Fig. 10. ON-OFF ratio for a three-bit-stream probe. The input probe energies are 0.01 pJ and 0.1 pJ, and pulsewidth is 1 ps. The input pump pulse energies are 0.1 pJ and 1 pJ.

In the experiments reported so far, 100 – 6.3 Gbit/s (Kawanishi et al., 1994; Uchiyama et al., 1998), 200 – 6.3 Gbit/s (Morioka et al., 1996), 40 – 10 Gbit/s (Tomkos et al., 1999), and 100 – 10 Gbit/s (Kirita et al., 1998) demultiplexing were performed. In our simulation/ modeling, we have considered the nonlinear effects, CD, CH and SHB with the recovery times of 200 ps, 700 fs, and 60 fs, respectively (Hong et al., 1996). Because, we assumed a probe pulse repetition rate of 250 Gbit/s, which is much faster than the recovery time of the CD, the CD caused by the probe pulses remains when the following probe pulses are injected into the SOA. Therefore, the pattern effect may arise and deteriorate the DEMUX operation for the multi-bit probe pulses (Saleh & Habbab, 1990). Here, we have considered the pattern effect of the probe bits because the different number of probe pulses is injected between the consecutive pump pulses depends on the bit pattern. Figure 11 shows the schematic diagram for the simulation of an optical DEMUX to investigate the pattern effect appearing at the DEMUX signals. The repetition rate and the pulsewidth of the input probe pulse are set to be 250 Gbit/s and 1 ps, respectively. We have simulated the FWM signals for the case that the different number of the probe pulses, n-1 are injected before the demultiplexed signal is extracted. More number of probe pulses reduce the DEMUX (FWM) signals as shown by the dashed line (where, n = 30).

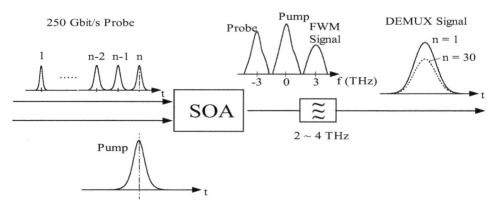

Fig. 11. Schematic diagram for the simulation of pattern effect on all-optical DEMUX operation. The input probe pulse repetition rate and pulsewidth are 250 Gbit/s and 1 ps, respectively. The different number of the probe pulses n-1 are injected before the demultiplexed signal is extracted.

Figure 12(a) shows an example of the pattern effect on the DEMUX signal waveforms. The input pump and probe pulse energies are 0.1 pJ (E_p) and 0.1 pJ (E_q), respectively. With increase the number of probe pulses, the FWM signal peak power decreases. The reduction in the peak power amounts to 7.4% for 30 probe pulses, while the waveforms remain unaffected. Fig. 12(b) shows the FWM signal energy versus the number of probe pulses. The closed circles show the calculated results and the solid line shows the fitted curve under the following approximation (Das, 2000; Das et al., 2001). The FWM signal is generated through the modulations in the refractive index and gain in the active region of SOAs. The modulation depths are proportional to both the carrier density and photon density because the modulation is created by the stimulated emission. Therefore, the FWM signal may also be proportional to the carrier density and photon density. The rate equation that describes the carrier density N in the active region is given by

$$\frac{dN}{dt} = \beta(N_0 - N) - aPN \tag{17}$$

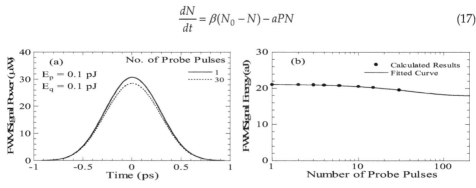

Fig. 12. (a) Pattern effect on the DEMUX signal waveforms and (b) FWM signal energy against number probe pulses. The input pump and probe energies are 0.1 pJ and 0.1 pJ, respectively.

where, P is the photon density, $\beta\ (= 1/\tau_s)$ is the recovery rate of the carrier density, N_0 is the carrier density at $t = 0$ (before the injection of probe pulse), and a is the coefficient of the stimulated emission. At $t = 0$, $N = N_0$, then the solution of equation (17) is as follows:

$$N = \frac{\beta N_0}{aP + \beta}\left[1 + \frac{aP}{\beta}\exp\{-(aP + \beta)t\}\right] \tag{18}$$

The repetition rate of the probe pulse is 250 GHz. Thus, we assume that the light with a constant photon density P is injected. The duration of the probe bits is given by $t = n/(2.5 \times 10^{11})$. Because, the FWM signal intensity is proportional to the carrier density N, the FWM signal intensity S_{FWM} can be expressed as follows.

$$S_{FWM}(t) = A\left[1 + B\exp(-t/\tau)\right] \tag{19}$$

where, $A = \beta N_0 \tau$, $B = aP/\beta$ and $\tau = 1/(aP + \beta)$. In Fig. 12(b) the solid line shows the fitted curve using the equation (19) with the parameters, A of 18.0 aJ, B of 0.178, and τ of 172 ps. Here, $A(1+B)$ corresponds to the maximum FWM signal intensity at $t = 0$, B is the constant representing the decrease in the FWM signal intensity caused by the probe pulses and τ is the effective recovery time of the carrier density depending on the input probe intensity. From equation (20), we obtained that the maximum fluctuation reaches to ~15% for the infinite number of probe pulse train.

$$\frac{S_{FWM}(0) - S_{FWM}(\infty)}{S_{FWM}(0)} = \frac{B}{1 + B} \tag{20}$$

Figure 13 shows another example of the pattern effects on the DEMUX signal for an input pump energy of 1 pJ and a probe energy of 0.01 pJ. In this case, the FWM signal intensities are stronger than the results shown in Fig. 12, because the probe energy is 10 times lower and the pump energy is 10 times as stronger than in Fig. 12. The FWM signal energy decreases by only 0.03% for 30 probe pulses. We have obtained FWM signal energy A of 62.0 aJ, B of 0.012, and τ of 200 ps from the fitted curve of Fig. 13(b). In this case, the FWM signal energy reduces only by 1.14% for the infinite number of probe pulses. We believe that such a small fluctuation is not an obstacle for the practical application. Although the results are not shown here, another set of the calculations are carried out shown in the Fig. 13, where the input pump and probe energy are 1 pJ and 0.1 pJ, respectively. The input pump energy is 10 times stronger than that of Fig. 12. The FWM signal intensities were about 100 times stronger than the results shown in Fig. 12. The FWM signal peak power decreases by less than 3% for 30 probe pulses. From the fitting to the calculations, we have obtained that A is 0.565 fJ, B is 0.09, and τ is 175 ps. Therefore, the FWM signal energy reduces by 9% for the infinite number of probe pulses in this condition. From these results, we can conclude that the intensity fluctuation of the FWM signal can be decreased by using the strong pump pulses or/and the weak input probe pulses.

The effective recovery time of the carrier density τ is defined as $\tau = 1/(aP + \beta)$. Therefore, in the weak limit of the probe pulses, τ should correspond to the carrier recovery time τ_s. In the weak probe case as shown in Fig. 13(b), τ is 200 ps and agrees with τ_s. In the case of the strong probe pulses, τ becomes short due to the stimulated emission caused by the probe pulses. For a strong probe pulse energy of 0.1 pJ, τ becomes smaller and they are 172 ps and 175 ps for pump pulse energies of 1 pJ and 0.1 pJ, respectively. These results support our assumptions as mentioned above.

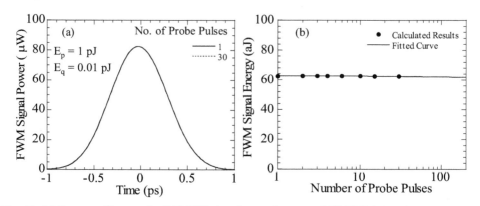

Fig. 13. (a) Pattern effect on the DEMUX signal waveforms and (b) FWM signal energy against number probe pulses. For this case, the input pump and probe energies are 1 pJ and 0.01 pJ, respectively.

Figure 14 shows an example of temporal waveforms of demuliplexed signals from time multiplexed signals by repetitive pump pulses. The input pump and probe pulse energies are 1 pJ and 0.01 pJ, respectively. The probe pulses are with a pulsewidth of 1 ps, sech^2 shape and have a repetition rate of 250 GHz. The pump pulses are with a pulsewidth of 1 ps, sech^2 shape and have a repetition rate of 62.5 GHz. Therefore, the 62.5 Gbit/s demultiplexed signals are selected once every four bits from the 250 Gbit/s signals. The FWM signal power is decreased by the strong input pump power due to the gain saturation and reaches to the constant value which is ~23% of the FWM signal power among the solitary pulses. There will be no pattern effect due to the gain saturation caused by the pump power, because the pump pulses are injected continuously. In this particular case with a low probe pulse energy of 0.01 pJ, the pattern effect caused by the probe pulse is expected to be very small as shown in Fig. 13.

One of the most important effects we have not included in this modeling is an amplified spontaneous emission (ASE) noise which is generated in SOAs. However, a number of literature emphasized the importance of the ASE noise. Summerfield and Tucker (Summerfield & Tucker, 1995) defined and measured the noise figure of an optical frequency converter based on FWM in an SOA. Diez et al., (Diez et al., 1997) defined the signal-to-background ratio (SBR) and investigated that different optimization criteria than for continuous waves apply as far as pulsed FWM applications concerned. Diez et al., (Diez et al., 1999) also reported a strong dependence of both conversion efficiency and SBR on pulsewidth and bit rate. This behavior has been attributed to the dynamics of the ASE, which is the main source of noise in an SOA.

Although the level of ASE strongly depends on the SOA structure and the operation conditions of SOAs, we have roughly compared with our calculated results and the ASE level on the assumption that the ASE level is – 40 dBm/nm (Diez *et al.*, 1997). In Fig. 6(a), the FWM signal is directly compared with ASE level. The FWM signal is about 10 dB greater than the ASE level when the pump pulsewidth is 1 ps. In Fig. 6(b), the ASE level becomes ~ 0.5 μW if we use a filter with 2 THz bandwidth. We can observe very clearly the FWM signal in the time domain. In Fig. 7, the ASE level becomes ~ 8 aJ if we use a filtering with 2 THz bandwidth and select a time slot of 16 ps (i.e., consider a 62.5 GHz repetition rate). Therefore, except for the conditions of a 0.1 pJ pump pulse and a 0.01 pJ probe pulse, the energy of FWM signal is greater than the ASE level. For more detailed comparison, it is recommended to take into account the ASE effect and its dynamic characteristics in the simulation.

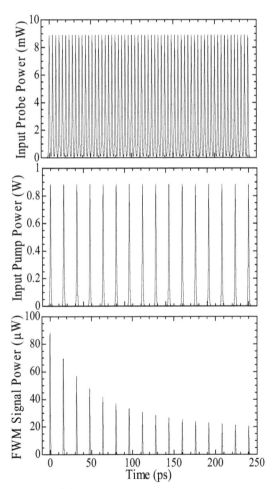

Fig. 14. DEMUX signal characteristics for the repetitive pump pulses. The input pump and probe pulse energies are 1 pJ and 0.01 pJ, respectively. Here, top figure is the input probe pulses, middle figure is the input pump pulses and bottom figure is the generated FWM signal pulses.

5. Conclusion

We have presented a detail analysis of all-optical DEMUX based on FWM in SOAs by solving the modified nonlinear Schrödinger equation using the FD-BPM. From this analysis, it was clarified that the optimization of the input pump pulsewidth is crucial to achieve a high ON-OFF ratio. We have obtained an optimum input pump pulsewidth of 1 ~ 3 ps for 1 ps, 250 Gbit/s input probe pulses. The shorter limit of the pulsewidth is due to detuning between the pump and probe frequency, which is determined by the gain bandwidth of the SOA. In order to realize faster DEMUX operation, SOAs with broader gain bandwidth are required. We have also simulated pattern effects in the FWM signal. When the number of input probe pulses increases, the FWM signal power decreases, however; the FWM signal waveforms remain unaffected. The peak power fluctuation of the FWM signal can be reduced by using the strong pump pulses and/or weak probe pulses. The energy fluctuation of the FWM signal decreases to less than 1% for a 30-bits, 250-Gbit/s input probe pulse train with a pulse energy of 0.01 pJ. This small fluctuation should not disturb the practical DEMUX operation. We also confirmed the DEMUX from time multiplexed signals by repetitive pump pulses. The strong energy pump pulses decrease the FWM signal intensity, however, there is no pattern effect due to gain saturation, because the pump pulses are injected continuously.

6. Acknowledgments

The authors would like to thank Dr. T. Kawazoe and Mr. Y. Yamayoshi for their helpful contribution to this work.

7. References

Agrawal, G. P. (1989). *Nonlinear Fiber Optics*. Academic Press, San Diego, Calif. ISBN 0-12-045142-5.

Agrawal, G. P. & Olsson, N. A. (1989). Self-phase modulation and spectral broadening of optical pulses in semiconductor laser and amplifiers. *IEEE J. Quantum Electron.*, vol. 25, pp. 2297-2306, ISSN 0018–9197.

Aghajanpour, H.; Ahmadi, V. & Razaghi, M. (2009). Ultra-short optical pulse shaping using semiconductor optical amplifier. *Optics & Laser Technology*, vol. 41, pp. 654-658, ISSN 0030-3992.

Brigham, E. Oran (1988). *The Fast Fourier Transform and Its Applications*. Englewood Cliffs, N.J.: Prentice-Hall Inc. ISBN 0-13-307505-2.

Buxens, A.; Poulsen, H. N., Clausen, A. T., & Jeppesen, P. (2000). All-optical OTDM-to-WDM signal-format translation and OTDM add-drop functionality using bidirectional four wave mixing in semiconductor optical amplifier. *Electron. Lett.*, vol. 36, pp. 156-158, ISSN 0013-5194.

Chung, Y. & Dagli, N. (1990). An Assessment of finite difference beam propagation method. *IEEE J. Quantum Electron.*, vol. 26, pp. 1335-39, ISSN 0018–9197.

Connelly, M. J.; Barry, L. P., Kennedy, B. F. & Ried, D. A. (2008). Numerical analysis of four-wave mixing between picosecond mode-locked laser pulses in a tensile-strained bulk SOA. *Optical and Quantum Electronics*, vol. 40, pp. 411-418, ISSN 1572-817X.

Conte, S. D. & Boor, Carl de (1980). *Elementary Numerical Analysis: An Algorithmic Approach,* Third Edition, McGraw-Hill Book Company Co., Singapore. ISBN 0070124477.

Das, N. K. (2000). *Numerical simulations of four-wave mixing among short optical pulses in semiconductor optical amplifiers by the beam propagation method.* PhD dissertation, Yamagata University, Japan.

Das, N. K.; Yamayoshi, Y. & Kawaguchi, H. (2000). Analysis of basic four-wave mixing characteristics in a semiconductor optical amplifier by beam propagation method. *IEEE J. Quantum Electron.,* vol. 36, pp. 1184-1192, ISSN 0018–9197.

Das, N. K. & Karmakar, N. C. (2008). Nonlinear propagation and wave mixing characteristics of pulses in semiconductor optical amplifiers. *Microwave and Optical Technology Letters,* vol. 50, pp. 1223-1227, ISSN 0895-2477.

Das, N. K.; Kawaguchi, H. & Alameh, K. (2011). *Advances in Optical Amplifiers,* Paul Urquhart (Ed.), "Ch. 6: Impact of Pump-Probe Time Delay on the Four-Wave Mixing Conversion Efficiency in SOAs", InTech, Austria. ISBN 978-953-307-186-2.

Das, N. K.; Karmakar, N. C., Yamayoshi, Y. & Kawaguchi, H. (2007). Four-wave mixing characteristics in SOAs with optimum time-delays between pump and probe pulses," *Microwave and Optical Technology Letters,* vol. 49, pp. 1182-1185, ISSN 0895-2477.

Das, N. K.; Yamayoshi, Y., Kawazoe, T. & Kawaguchi, H. (2001). Analysis of optical DEMUX characteristics based on four-wave mixing in semiconductor optical amplifiers. *IEEE/OSA J. Lightwave Technol.,* vol. 19, pp. 237-246, ISSN 0733-8724.

Das, N. K.; Kawazoe, T., Yamayoshi, Y. & Kawaguchi, H. (2001). Analysis of optical phase-conjugate characteristics of picosecond four-wave mixing signals in semiconductor optical amplifiers. *IEEE J. Quantum Electron.,* vol. 37, pp. 55-62, ISSN 0018–9197.

Das, N. K.; Karmakar, N. C., Yamayoshi, Y. & Kawaguchi, H. (2005). Four-wave mixing characteristics among short optical pulses in semiconductor optical amplifiers with optimum time-delays, *Proceedings of the 18th Annual Meeting of the IEEE Lasers and Electro-Optics Society 2005 (IEEE-LEOS2005),* pp. 127-128, ISBN 0-7803-9217-5, Sydney, NSW, Australia, October 2005, IEEE Press (USA).

Dienes, A.; Heritage, J. P., Jasti, C. & Hong, M. Y. (1996). Femtosecond optical pulse amplification in saturated media. *J. Opt. Soc. Am. B,* vol. 13, pp. 725-734, ISSN 0740-3224.

Diez, S.; Schmidt, C., Ludwig, R., Weber, H. G., Obermann, K., Kindt, S., Koltchanov, I. & Petermann, K. (1997). Four-wave mixing in semiconductor optical amplifiers for frequency conversion and fast optical switching. *IEEE J. Sel. Top. Quantum Electron.,* vol. 3, pp. 1131-1145, ISSN 1939-1404.

Diez, S.; Mecozzi, A. & Mørk, J. (1999). Bit rate and pulse width dependence of four-wave mixing of short pulses in semiconductor optical amplifiers. *Opt. Lett.,* vol. 24, pp. 1675-1677, ISSN: 0146-9592.

Eiselt, M. (1995). Optimum pump pulse selection for demultiplexer application of four-wave mixing in semiconductor laser amplifiers. *IEEE Photon. Technol. Lett.,* vol. 7, pp. 1312-1314, ISSN 1041-1135.

Gingrich, H. S.; Chumney, D. R., Sun, S.-Z., Hersee, S. D., Lester, L. F. & Brueck, S. R. (1997). Broadly tunable external cavity laser diodes with staggered thickness multiple quantum wells. *IEEE Photon. Technol. Lett.,* vol. 9, pp. 155-157, ISSN 1041-1135.

Hong, M. Y.; Chang, Y. H., Dienes, A., Heritage, J. P., Delfyett, P. J., Dijaili, Sol & Patterson, F. G. (1996). Femtosecond self- and cross-phase modulation in semiconductor laser amplifiers. *IEEE J. Sel. Top Quantum Electron.*, vol. 2, pp. 523-539, ISSN 1939-1404.

Inoue, J. & Kawaguchi, H. (1998). Highly nondegenerate four-wave mixing among sub-picosecond optical pulses in a semiconductor optical amplifier," *IEEE Photon. Technol. Lett.*, vol. 10, pp. 349–351, ISSN 1041-1135.

Kikuchi, K. & Matsumura, K. (1998). Transmission of 2-ps optical pulses at 1550 nm over 40-km standard fiber using midspan optical phase conjugation in semiconductor optical amplifiers. *IEEE Photon. Technol. Lett.*, vol. 10, pp. 1410-1412, ISSN 1041-1135.

Kawanishi, S.; Okamoto, K., Ishii, M., Kamatani, T., Takara, H., & Uchiyama, K. (1997). All-optical time-division-multiplexing on four-wave mixing in a travelling-wave semiconductor laser amplifier. *Electron. Lett.*, vol. 33, pp. 976-977, ISSN 0013-5194.

Kawanishi, S.; Morioka, T., Kamatani, O., Takara, H., Jacob, J.M., & Saruwatari, M. (1994). 100 Gbit/s all-optical demultiplexing using four-wave mixing in a travelling wave laser diode amplifier. *Electron. Lett.*, vol. 30, pp. 981-982, ISSN 0013-5194.

Kirita, H.; Hashimoto, Y., & Yokoyama, H. (1998). All-optical signal processing at over 100 Gbit/s with nonlinear effects in semiconductor lasers. *Tech. Dig. Int'l. Trop. Workshop on Contemporary Technologies (CPT '98)*, Paper Pc-14, Tokyo, Japan.

Koltchanov, I.; Kindt, S., Petermann, K., Diez, S., Ludwig, R., Schnabel, R., & Weber, H. G. (1996). Gain dispersion and saturation effects in four-wave mixing in semiconductor laser amplifiers. *IEEE J. Quantum Electron.*, vol. 32, pp. 712-720, ISSN 0018-9197.

Leuthold, J.; Mayer, M., Eckner, J., Guekos, G., Melchior, H. and Zellweger, Ch. (2000). Material gain of bulk 1.55 μm InGaAsP/InP semiconductor optical amplifiers approximated by a polynomial model. *J. Appl. Phys.*, vol. 87, pp. 618-620, ISSN.

Mecozzi, A. & Mørk, J. (1997). Saturation effects in nondegenerate four-wave mixing between short optical pulses in semiconductor laser amplifiers. *IEEE J. Sel. Top Quantum Electron.*, vol. 3, pp. 1190-1207, ISSN 1939-1404.

Mecozzi, A.; D'Ottavi, A., Iannone, E., & Spano, P. (1995). Four-wave mixing in travelling-wave semiconductor amplifiers. *IEEE J. Quantum Electron.*, vol. 31, pp. 689-699, ISSN 0018-9197.

Mikami, O.; Noguchi, Y., Yasaka, H., Magari, K. & Kondo, S. (1991). Emission spectral width broadening for InGaAsP/InP superluminescent diodes. *IEE Proc. J Optoelectron.*, vol. 138, pp. 133-137, ISSN: 0267-3932.

Mørk, J. & Mecozzi, A. (1997). Theory of nondegenerate four-wave mixing between pulses in a semiconductor waveguide. *IEEE J. Quantum Electron.*, vol. 33, pp. 545-555, ISSN 0018-9197.

Morioka, T.; Takara, H., Kawanishi, S., Uchiyama, K., & Saruwatari, M. (1996). Polarisation-independent demultiplexing up to 200 Gb/s using four-wave mixing in a semiconductor laser amplifier. *Electron. Lett.*, vol. 32, pp. 840-841, ISSN 0013-5194.

Okamoto, K. (1992). *Theory of Optical Waveguides*, Corona Publishing Co., Tokyo; Ch. 7 (in Japanese). ISBN 4-339-00602-5.

Razaghi, M.; Ahmadi, A., Connelly, M. J. & Madanifar, K. A. (2009a). Numerical modelling of sub-picosecond counter propagating pulses in semiconductor optical amplifiers. *Proceedings of the 9th International Conference on Numerical Simulation of Optoelectronic*

Devices 2009 (NUSOD' 09), pp. 59-60, ISBN 978-1-4244-4180-8, GIST, Gwangju, South Korea, September 2009, IEEE Press (USA).

Razaghi, M.; Ahmadi, A., & Connelly, M. J. (2009b). Comprehensive finite-difference time-dependent beam propagation model of counter propagating picosecond pulses in a semiconductor optical amplifier. *IEEE/OSA J. Lightwave Technol.*, vol. 27, pp. 3162-3174, ISSN 0733-8724.

Saleh, A. A. A. & Habbab, I. M. I. (1990). Effects of semiconductor-optical-amplifier nonlinearity on the performance of high-speed intensity-modulation lightwave systems. *IEEE Trans. Commun.*, vol. 38, pp. 839-846, ISSN 0090-6778.

Seki, K.; Kamiya, T. and Yanai, H. (1981). Effect of waveguiding properties on the axial mode competition in stripe-geometry semiconductor lasers. *IEEE J. Quantum Electron.*, vol. 17, pp. 706-713, ISSN 0018–9197.

Shtaif, M. & Eisenstein, G. (1995). Analytical solution of wave mixing between short optical pulses in semiconductor optical amplifier. *Appl Phys. Lett.* 66, pp. 1458-1460, ISSN 0003-6951.

Shtaif, M.; Nagar, R. & Eisenstein, G. (1995). Four-wave mixing among short optical pulses in semiconductor optical amplifiers. *IEEE Photon Technol. Lett.* 7, pp. 1001-1003. ISSN 1041-1135.

Shtaif, M. & Eisenstein, G. (1996). Calculation of bit error rates in all-optical signal processing applications exploiting nondegenerate four-wave mixing in semiconductor optical amplifiers. *IEEE/OSA J. Lightwave. Technol.*, vol. 14, pp. 2069-2077, ISSN 0733-8724.

Summerfield, M. A. & Tucker, S. R. (1995). Noise figure and conversion efficiency of four-wave mixing in semiconductor optical amplifiers. *Electron. Lett.*, vol. 31, pp. 1159-1160, ISSN 0013-5194.

Sauter, E. G. (1996). *Nonlinear Optics.* John Wiley & Sons, Inc. New York. ISBN 0-471-14860-1.

Tang, J. M. & Shore, K. A. (1998). Influence of probe depletion and cross-gain modulation on four-wave mixing of picosecond optical pulses in semiconductor optical amplifiers," *IEEE Photon. Technol. Lett.*, vol. 10, pp. 1563-1565, ISSN 1041-1135.

Tang, J. M. & Shore, K. A. (1999). Active picosecond optical phase compression in semiconductor optical amplifiers. *IEEE J. Quantum. Electron.*, vol. 35, pp. 93-100, ISSN 0018–9197.

Tang, J. M. & Shore, K. A. (1999). Characteristics of optical phase conjugation of picosecond pulses in semiconductor optical amplifiers. *IEEE J. Quantum. Electron.*, vol. 35, pp. 1032-1040, ISSN 0018–9197.

Tang, J. M.; Spencer, P. S. & Shore, K. A. (1998). The influence of gain compression on picosecond optical pulses in semiconductor optical amplifiers. *J. Mod. Opt.*, vol. 45, pp. 1211-1218, ISSN: 0950-0340.

Tomkos, I.; Zacharopoulos, I., Syvridis, D., Calvani, R., Cisternino, F., & Riccardi, E. (1999). All-optical demultiplexing/shifting of 40-Gb/s OTDM optical signal using dual-pump wave mixing in bulk semiconductor optical amplifier. *IEEE Photon. Technol. Lett.*, vol. 11, pp. 1464-1466, ISSN 1041-1135.

Uchiyama, K.; Kawanishi, S., & Saruwatari, M. (1998). 100-Gb/s multiple-channel output optical OTDM demultiplexing using multichannel four-wave mixing in a semiconductor optical amplifier. *IEEE Photon. Technol. Lett.*, vol. 10, pp. 890-892, ISSN 1041-1135.

Vahala, K. J.; Zhou, J., Geraghty, D., Lee, R., Newkirk, M., & Miller, B. (1996). Four-wave mixing in semiconductor travelling-wave amplifiers for wavelength conversion in all-optical networks, in T. P. Lee ed. *Current Trends in Optical Amplifiers and Their Applications.* World Scientific, Singapore.

Xie, C.; Ye, P. & Lin, J. (1999). Four-wave mixing between short optical pulses in semiconductor optical amplifiers with the consideration of fast gain saturation. *IEEE Photon. Technol. Lett.*, vol. 11, pp. 560-562, ISSN 1041-1135.

Yariv, A. (1991). *Optical Electronics*, 4th Edition, Saunders College Publishing, San Diego. ISBN 0-03-053239-6.

8

Realization of HDWDM Transmission System with the Minimum Allowable Channel Interval

Jurgis Porins, Vjaceslavs Bobrovs and Girts Ivanovs
Riga Technical University, Institute of Telecommunications
Latvia

1. Introduction

Nowadays, a skyrocketing growth is observed worldwide in the bit rates of transmitted information, which is associated with development of broadband information transmission types. The annual global internet protocol (IP) traffic will exceed half a Zettabyte in four years. At just under 44 Exabytes per month, the annual run rate of traffic in the late 2012 will be 522 Exabytes per year. Driven by high-definition video and high-speed broadband penetration, the consumer of IP traffic will bolster the overall IP growth rate so that it sustains a steady growth rate through 2012, growing at a compound annual growth rate (CAGR) of 46 percent (see Fig. 1) [Cisco Systems, 2008].

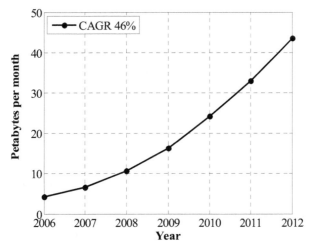

Fig. 1. Global IP Traffic Forecast (2006–2012) [Cisco Systems, 2008].

In turn, to provide high-quality transmission it is necessary to develop the next generation optical networks (NGONs) that would transmit properly huge volumes of information. The optical transmission systems from the very outset have been able to offer new possibilities for solving problems of ever increasing urgency that are dictated by the need for frequency bands and transmission speed. Such networks have become one of the most important components in the telecommunication hierarchy, whose integration with standard network services and applications promotes rapid evolution of fiber optics and its wide implementation into all telecommunication branches (see Table 1 [McGloin & Reid., 2010]).

Transmission Network	CORE	METRO	ACCESS	CROSS CONNECTIONS
Length	> 100 km	10 km	Approx. 20 km (ITU recommendation), normally < 10 km	< 100 m
Laser type	DFB	DFB,VCSEL	DFB or Fabry-Perot	VCSEL
Wavelength	1550 nm central ± 30 nm	1310 nm; 1550 nm	1310 nm; 1490 nm; 1550 nm	850 nm; 1310 nm
Modulation scheme	External	External or internal	Internal	Internal
Bit rate	10 Gbit/s	10 Gbit/s	< 2.5 Gbit/s	Depending on protocol type
Multiplexing scheme	DWDM or CWDM	CWDM	WDM	Depending on protocol type

Table 1. Evolution of fiber optics and its wide implementation into all telecommunication branches

Currently, many research topics in the field of optical transmission systems (mostly grounded on novel modulation techniques) are focused on increasing the total data transmission speed of an individual optical fiber [Abbou et al., 2008, Bhamber et al., 2007, Bobrovs et al., 2008]. An alternative – but equally valid – approach to increasing the data transmission is to decrease the wavelength division multiplexing (WDM) channel spacing to high-dense dimensions while keeping the existing data transmission speed for an exact channel [Ozoliņš et al., 2011, Bobrovs et al., 2009].

High performance optical filters make the groundwork for realization of high-speed high-density WDM (HDWDM) transmission systems [Pfennigbauer & Winzer, 2006]. High channel spacing and data transmission rate set strict requirements for HDWDM filter characteristics, so any imperfections in their parameters, such as amplitude and phase responses, could become critical. The low channel separation from adjacent channels is one of these imperfections in optical filter parameters [Agrawal, 2001, Ozoliņš et al., 2009].

2. Implementation of HDWDM transmission system

Due to these rapidly growing capacity requirements for long-haul transmission, the optical wavelength division multiplexing systems are advancing into high data transmission rate and dense channel spacing to utilize the available bandwidth more effectively. In order to maximize the system capacity and to minimize the performance degradation caused by transmission impairments the system investigation and optimization are very important. To increase the spectral efficiency is important for building efficient HDWDM transmission systems, since this allows the optical infrastructure to be shared among many wavelengths. This approach reduces the cost per transmitted information bit in a fully loaded and optimized transmission system.

2.1 Selection of HDWDM main components

The complexity of a system's design in optical communications can be seen as the result of a large number of components with different parameters and operational states. The description of the interaction between the optical signal and transmission disturbances is a multi-dimensional issue, whose solution depends on the relation between different system parameters. The right approach to the optimization of system settings and derivation of design rules must take into account the interaction of effects which take place in each component. In this section, the system components needed for realization of an HDWDM transmission system are described.

The role and realization of an optical transmitter become important with increased channel data rates in the system. While the optical transmitters at lower channel data rates are less complex and easier to realize by direct modulation of a laser diode, the realization becomes more complex with the increasing channel data rate, thus raising the requirements on electrical and optical components of the optical transmitter. The conventional optical transmitter employs the amplitude/intensity modulation (AM, IM) of the laser light (better known as on-off keying (OOK)), because different signal levels for marks and spaces are characterized by the presence of optical power. The amplitude modulation can be realized by direct or external modulation of the laser diode. For the realization of transmission systems with channel data rates larger than 2.5 Gbit/s, the external modulation presents a better solution, because the impact of laser internal chirp on optical signal can be reduced efficiently, but, on the other hand, the complexity of optical transmitters increases.

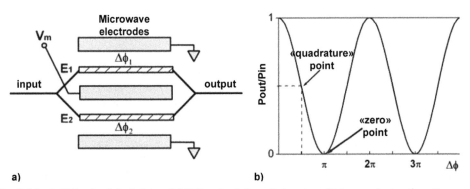

Fig. 2. Mach-Zehnder Modulator (MZM) principles: a) structure b) transmission function [Kaminow et al., 2009].

External modulation can be realized with a $LiNbO_3$-based Mach-Zehnder modulator (MZM) (see Fig. 2) [Kaminow et al., 2009]. The operational principles of MZMs are based on the electro-optic effect, which is characterized by variation in the applied electrical field causing changes of the refractive index in the modulator arms. The variation of the refractive index in the modulator arms induces a change of material propagation constant β, resulting in different phases in both modulator arms. The input optical signal is divided by a 3-dB coupler into two equal parts – in lower and upper arm of the MZM. The external modulator is driven by an electrical signal with corresponding data rate. Depending on the electrical driving signal, different transmission speeds can be realized. If no electrical field is applied, both signals arrive at the same time (in-phase) at the MZM output and interfere constructively. If an electrical filed is applied, signals in different arms are shifted in phase relative to each other. Depending on the phase difference between the MZM arms, the signals can interfere constructively or destructively, resulting in an amplitude modulation of the modulator input signal. In this signal generation method, the laser source acts as a continuous wave (CW) pump. In conventional systems, the CW pumps are realized with distributed feedback laser (DFB) (the most important and widely used single mode laser type for the 1550 nm region). DFB lasers are realized by the implementation of through a Bragg's grating structure inside the cavity between the reflecting surfaces of a laser [Voges & Peteramann, 2002]. The main characteristics of the DFB lasers are high side-mode suppression ratios (> 50 dB) enabling stable single-mode operation, a small spectral line width (0.8...50 MHz) and large output optical power (10...40 mW) (see Fig. 3) [Funabashi, 2001].

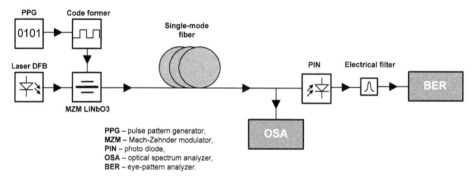

Fig. 3. Simplified fiber optical transmission system.

After the MZM, such a signal is sent directly to a transmission medium, where optical pulses are propagating over different distances of a single-mode fiber (SMF). For compensation of losses in the fiber and in optical components it is necessary to use the technique for amplifying optical signals. The optical amplifiers represent one of the crucial components in an optical transmission system. Despite the minimum attenuation at 1550 nm, fiber losses significantly limit the transmission performance with increased transmission distance. Optical amplification can be realized using different amplifier concepts and mechanisms, e.g. semiconductor optical amplifiers (SOA), rare-earth (erbium, holmium, thulium, and samarium) doped fiber amplifiers, or, more recently, Raman amplifiers [Kaminow et al., 2008, Binh, 2008]. All these amplifier types are based upon different physical mechanisms resulting in different device characteristics and application areas. The rare-earth-doped fiber amplifiers provide optical amplification in the wavelength region from 500 to 3500 nm. The most important representative of these amplifier types is the erbium-doped fiber amplifier (EDFA), which is widely used today in optical transmission systems since it provides efficient optical amplification in the 1550 nm region. The EDFAs present the state-of-the-art technology in conventional optical transmission systems, and they can be used as in-line amplifiers (placed every 30-80 km), power boosters (amplifiers at the transmitter side) or pre-amplifiers (amplifiers in front of the receiver) independently of the channel bit rate in the system [Thyagarajan & Ghatak, 2007].

After transmission through the optical fiber, a multiwavelength optical signal needs to be separated in individual channels. This is realized through implementation of band-pass filters (BPFs), which transmit optical power within a definite wavelength window only, and reflect or absorb the rest. In the case of a single-channel transmission the function of an optical BPF is to separate the channel information from the noise which has been added, e.g., by optical amplifiers. This noise is generally broadband, and can often be described as quasi-white: it has a constant level in the power spectrum [Kashyap, 2010, Venghaus, 2006]. By applying a BPF to select the wavelength channel, the useful information is retained and most of the noise is filtered resulting in an improvement of the optical-signal-to-noise ratio (OSNR). Such a filter can also be used to select a particular channel in a HDWDM application from several channels that are transmitted in a common HDWDM transmission system [Azadeh, 2009, Venghaus, 2006]. The role of an optical receiver is to detect the transmitted signal by the opto-electrical transformation of the signal received by a photo-diode (e.g. PIN or APD). Furthermore, additional electrical equalization is performed together with electrical signal amplification enabling further signal-processing (e.g. clock-recovery) and performance evaluation (e.g. quality measurements).

In fiber optical transmission systems, the degradation effects can be categorized by the random noise and waveform distortion. For long-span HDWDM systems, signal waveform distortion can be generated by linear chromatic dispersion, polarization mode dispersion (PMD), nonlinear optical effects (NOE) in optical fibers, or their combination [Chen et al., 2006, Pan et al., 2010]. In high-speed (more than 2.5 Gbit/s) time division multiplexing (TDM) optical systems having short optical pulses and wide optical spectrum the effect of complex dispersion dominates in the system performance degradation. In multiwavelength WDM optical systems the inter-channel crosstalk originated by fiber nonlinearity, such as cross-phase modulation (XPM) and four-wave mixing (FWM), is a limiting factor. To maximize the WDM network capacity, the system's design and optimization have to take into account all the contributing factors - such as the channel data rate, transmission distance, signal optical power, fiber linear and nonlinear optical effects and, of course, the channel interval [Venghaus, 2006]. In a HDWDM system the last factor is the most important for a high-quality solution which depends directly on the optical filters.

2.2 Optical filters for HDWDM systems

The wavelength filters in optical transmission systems are a special subgroup of physical components defined in such a way that they select or modify parts of the signal spectrum. In fact, the optical wavelength filters are defined as referred to the modifications that they induce in the frequency spectrum. In electronic systems, relevant filters play a crucial role in numerous signal processing applications. Similarly, optical filters play an equally crucial role in the optical domain [Venghaus, 2006, Kaminow et al., 2008].

Multiplexing and de-multiplexing functions are performed by narrowband filters, cascaded and combined in various ways to achieve the desired result. The filters in optical HDWDM transmission systems are classified into the following types: notch filters, power equalization filters, all-pass filters and band-pass filters [Szodenyi, 2004].

As was said above in sub-section 2.1, band-pass filters (BPFs) transmit optical power within a definite wavelength window only, and reflect or absorb the rest. The bandwidth of an optical

BPF typically depends on the optical transmission system type [Venghaus, 2006]. For example, in HDWDM the sharpness of the optical BPF amplitude transfer function is of great importance, while in the coarse wavelength division multiplexing (CWDM) it is a minor factor because of a wide frequency interval between the adjacent channels. In these systems the major role is played only by the optical BPF bandwidth, and in the DWDM systems also the shape of the amplitude and phase transfer function should be taken into account. Although different kinds of filters are necessary in a HDWDM transmission system, BPFs are by far the most important, since they are prerequisite for add and drop, multiplex, interleave and routing functionalities which are essentials for a HDWDM transmission system realization [Agrawal, 2001].

Travelling through a multiple optical BPF, the optical signal experiences spectral narrowing due to temperature instability of filtering devices and to central frequency fluctuations of light sources, which could be the main factor of degradation in future transmission systems. Therefore, it is necessary to find out the minimal filter's full width half maximum (FWHM) which ensures appropriate quality of transmitted data signals. Still, the filter bandwidth is not the exclusive parameter of which we need to be aware: the phase transfer function of optical band-pass filters is also of great importance for transmitting information via HDWDM transmission systems.

It is possible to employ three different transfer functions of the optical filter (see Fig. 4) for realization of HDWDM system schemes. These functions were chosen because with the Lorentzian optical filter's transfer function we can approximate: Fabry Perrot filters, micro-ring resonators; raised cosine filters: arrayed waveguide gratings with flat tops, diffraction gratings, and particular cases of thin film filters and fiber Bragg gratings (with apodization); supergaussian filters: arrayed waveguide gratings with supergaussian transfer function, and thin film filters with low refraction index modulation [Venghaus, 2006].

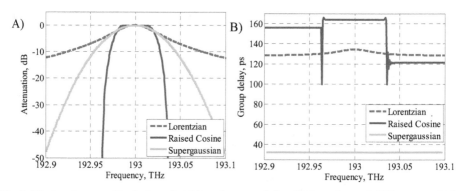

Fig. 4. First-order amplitude transfer (*a*) and group delay (*b*) functions of different optical filters shown in the inset (with FWHM bandwidth 0.4 nm or 50 GHz).
The graphs are obtained by/using OptSim simulation software.

As is seen from Fig. 4*b*, the greater group delay is for the Raised Cosine optical band-pass filter whose amplitude characteristics are the closest to an ideal filter's amplitude parameters. The ideal amplitude transfer function of a band-pass filter has an almost rectangular shape, providing a perfect transmission (without distortion) of the whole signal within the filter bandwidth, and cutting undesired signals out of the band [Venghaus, 2006].

3. Performance evaluation criteria

The right choice of the performance evaluation criteria for characterizing the optical transmission lines is one of the key issues in designing efficient high-speed systems. The evaluation criteria should provide precise determination and separation of dominant system limitations, which is crucial for suppressing the propagation disturbances. They should also provide comparison of experimental and numerical data to verify the numerical models applied.

3.1 Evaluation of bit error ratio (BER)

The bit error ratio (BER) evaluation is a straightforward and relatively simple method for performance evaluation based on counting the errors in the received bit streams. The error counting in a practical system with a transmission speed greater than 1 Gbit/s can be a long process, especially for realistically low BER values ($< 10^{-9}$). For investigation of performance of an optical transmission system by simulation, several effective statistical methods have been developed [Binh, 2009].

Conventional methods of Q-factor and hence BER calculation are based on the assumption of Gaussian noise distribution. However, new methods relying upon statistical processes with account for the distortion dynamics of optical fibers are necessary in order to include the common patterning effects.

The former statistical technique employs the expected maximization theory in which the *pdf* of the detected electrical signal is approximated as a mixture of multiple Gaussian distributions.

The latter technique is based on the generalized extreme values (GEV) theorem [Bierlaire et al., 2007, Markose & Alentorn, 2007]. Although this theorem is well-known in other fields (financial forecasting, meteorology, material engineering, see e.g. [Kotz & Nadarajah, 2000] to predict the probability of occurrence of extreme values, it has not yet been applied in optical communications.

Exactly as the BER set used in experimental transmission, the BER in the simulation of a particular HDWDM system configuration is calculated. In this case the BER is the ratio of the occurrence of errors (N_{error}) to the total number of transmitted bits (N_{total}) and given as:

$$BER = \frac{N_{error}}{N_{total}} \qquad (1)$$

The Monte-Carlo method offers a precise picture via the BER metric for all modulation formats and receiver types. The optical system configuration under a simulation test must include all the sources of impairments imposing on signal waveforms, including fiber impairments and amplified spontaneous emission (ASE) [Binh, 2009].

3.2 Optical signal-to-noise ration (OSNR)

The optical signal-to-noise ratio (OSNR) is a widely used evaluation criterion for characterizing the system performance in already deployed transmission lines. The optical noise created by transmission media and devices around an optical signal reduces the receiver's ability to correctly detect the signal. This effect can be suppressed by an optical filter placed before the optical receiver. Depending on the amplifier infrastructure used in a transmission system, the OSNR is proportional to the number of optical amplifiers and to the gain flatness of a single amplifier. This latter can be an especially critical issue in HDWDM systems, because of the gain non-uniformity in multi-span transmissions.

In practice, the OSNR can be found by measuring the signal power as the difference between the total power of the signal peak and the background noise; this latter, in turn, is determined by measuring the noise contributions on either side of the signal peak. However, it is difficult to separate measurements of the signal and noise power, because the latter in an optical channel is included in the signal power. The determination of this parameter in a HDWDM system can be made by interpolating it between the adjacent channels [Binh, 2009].

For a single EDFA with output power, P_{out}, the OSNR is given by [Jacobsen, 1994]:

$$OSNR = \frac{P_{out}}{N_{ASE}} = \frac{P_{out}}{(NFG_{op}-1)h\nu B_o},$$
(2)

where NF is the amplifier noise figure, G_{op} is the optical amplifier gain, $h\nu$ is the photon energy, and B_o is the optical bandwidth found by measurement. However, OSNR does not provide good estimation to the system performance when the main degrading sources involve the dynamic propagation effects such as dispersion and Kerr nonlinearity effects.

When addressing the value of an OSNR, it is important to define the optical measurement bandwidth over which the OSNR is calculated. To obtain this value, the signal power and noise power are derived by integrating all the frequency components over the bandwidth [Rongqing & O'Sullivan, 2009].

In practice, the signal and noise power values are usually measured directly, using the optical spectrum analyzer (OSA), which does the mathematics for the users and displays the resultant OSNR versus wavelength or frequency over a fixed resolution bandwidth. The value of $\Delta\lambda$ = 0.1 nm at 1550 nm, is widely used as a typical value for calculation of the OSNR.

4. HDWDM system experimental and simulation models

Our experimental transmission system (see Fig. 5) employs two optical channels with external intensity modulation (IM), and non-return-to-zero (NRZ) pulse shapes. The laser is always switched on and its light waves are modulated via the electro-optic MZM by output of data pulse sequence of a pulse pattern generator (PPG), using the principles of interferometer constructive and destructive interference to achieve ON and OFF of the light waves. After the MZ modulator the signal is sent to a single-mode fibre (SMF), where optical pulses are propagating over a 40 km distance. The utilized fiber has a large core effective area of 80 μm^2, attenuation a = 0.2 dB/km, nonlinear refractive coefficient n_k = $2.5 \cdot 10^{-20}$ cm/W and dispersion 16 ps/nm/km at the reference wavelength λ = 1550 nm [Kaminow et al., 2009].

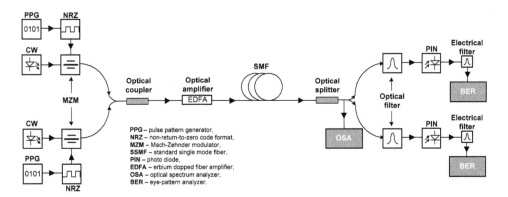

Fig. 5. The setup used for investigation of HDWDM transmission [Bobrovs et al., 2010].

At the fibre end each channel is optically filtered with an Anritsu Xtract tunable optical filter (see Fig. 6). An essential parameter of such a filter is its centering on the signal to be extracted. Its position has to be adjusted regarding the signal harmonics [Ivanovs et al., 2010].

The Anritsu Xtract tunable optical band-pass filter covers all transmission bands of a standard single mode optical fiber. The filter operates in the range of 1450-1650 nm, covering the E, S, C and L bands and, partially, the U band. The main drawback of this BPF is 6 dB insertion losses, which is a limiting factor in realization of high-speed HDWDM transmission systems for moderate distances without optical amplifiers.

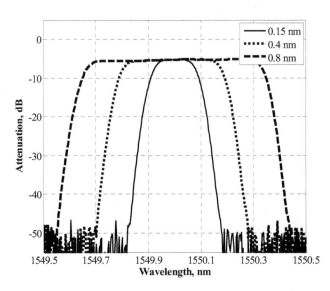

Fig. 6. The measured amplitude responses of the Anritsu Xtract tunable optical band-pass filter at different FWHM values.

To evaluate the output signal characteristics, an optical direct-detection receiver was used, with an electrical fourth-order Bessel–Thomson electrical filter having a 3 db bandwidth of 7.5 GHz. In practice, a 40 km span is preferred by most network providers since it allows a compromise between the system's costs and its performance [Binh, 2009].

For the performance evaluation and optimization of the experimental HDWDM system it is necessary to analyze the optical and electrical signal quality before MZM, after MZM and after SMF. The choice of arbitrary units on the Y-axis in the eye diagrams was purposeful – to make them more general in the cases when the plotted electrical quantity is current or voltage.

As a result, we have designed a HDWDM transmission system with a variable data transmission speed up to 12.5 Gbit/s, the channel interval up to 12.5 GHz and optical power up to 23 dBm. In Fig. 7 one can see 2.5 Gbit/s HDWDM transmission systems with 18.75 GHz and 25 GHz channel interval. As follows from the results, reducing the channel interval to 18.75 GHz gives rise to Kerr's effect, which degrades the 2.5 Gbit/s signal quality. The signal eye-pattern overlaps with the mask (see Fig. 7c), which means that the signal quality does not ensure the BER=10⁻⁹ value. To obtain a system with an appropriate BER we should reduce the data transmission speed or increase the channel interval. As can be seen from Fig. 5, the 25 GHz channel interval ensures a good signal quality, and the signal eye-pattern in this case does not overlap with the mask.

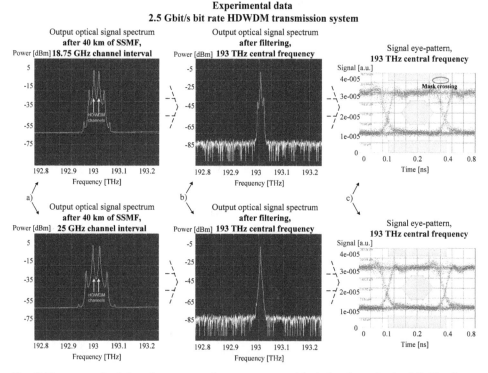

Fig. 7. Output optical signal spectra and eye-patterns with defined masks for 2.5 Gbit/s system: *a*) common optical spectra, *b*) signal optical spectrum after filtering, *c*) eye-pattern.

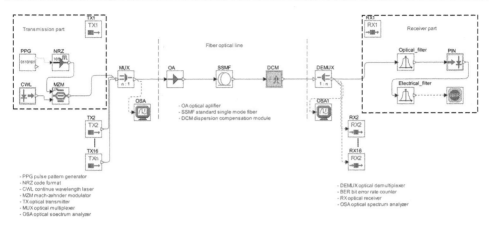

Fig. 8. Simulation model of HDWDM system.

In compliance with the experimental model we have created a simulation scheme (see Fig. 8) using OptSim software with the real parameters of all experimental devices. The accepted method of calculation is based on solving of a complex set of differential equations, taking into account optical and electrical noise as well as linear and nonlinear effects. We have used a model where signals are propagating as time domain samples over a selectable bandwidth (in our case, a bandwidth that contains all channels).

The Time Domain Split Step (TDSS) method was employed to simulate linear and nonlinear behavior for both optical and electrical components. The split step method is now used in all commercial simulation tools to perform the integration of a fiber propagation equation that can be written as [Binh, 2009]:

$$\frac{\partial A(t,z)}{\partial z} = \{L + N\} A(t,z) \tag{3}$$

Here $A(t,z)$ is the optical field, L is the linear operator that stands for dispersion and other linear effects, and N is the operator that is responsible for all nonlinear effects. The idea is to calculate the equation over small spans Δz of fiber by including either a linear or a nonlinear operator [Belai et al., 2006]. For instance, on the first span only linear effects are considered, on the second – only nonlinear, on the third – again only linear ones, and so on. Two ways of calculation are possible: frequency domain split step (FDSS) and the above-mentioned time domain split step (TDSS) method. These methods differ in how linear operator L is calculated: FDSS does it in a frequency domain, whereas TDSS – in the time domain, by calculating the convolution product in sampled time. The first method is easy to fulfill, but it may produce severe errors during computation. In our simulation we have employed the second method, TDSS, which, despite its complexity, ensures an effective and time-saving solution.

5. Results and discussions

Fig. 9 shows the spectral and eye diagrams in a simulative HDWDM communication system with 2.5 Gbit/s transmission speed per channel after a signal's detection. It is seen there that in the four-channel case the allowed interval is 12.5 GHz or 0.1 nm, with BER meeting the standard (< 10⁻⁹). If we raise the number of channels to 8, the output signal quality worsens; therefore, the channel interval should be raised up to 18.75 GHz (see Fig. 9). In turn, in the 16-channel case a successful transmission is possible only using 25 GHz (or 0.2 nm) channel spacing. This is the optimal channel interval, which allows multiplexing the signals in a HDWDM system with a channel number exceeding 16. Further increase in the channel number would not change the chosen frequency interval (see Fig. 9). Considering the 25 GHz frequency interval as the chosen one, it is possible to upgrade the existing WDM communication systems with a 2.5 Gbit/s transmission speed per channel without increasing this speed while decreasing the channel interval down to the estimated value and adding signals to the freed frequency range, thus realizing an HDWDM transmission system.

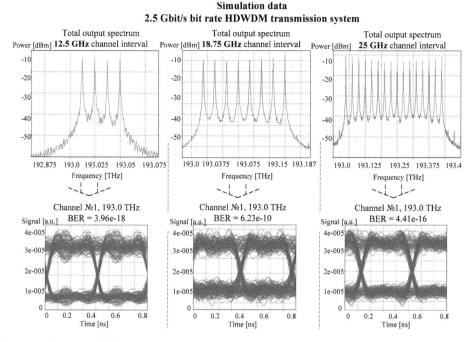

Simulation data
2.5 Gbit/s bit rate HDWDM transmission system

Fig. 9. 22.5 Gbit/s HDWDM communication system with 4/8/16 channels and 12.5/18.75/25 GHz frequency intervals. The output spectrum of optical signal is shown after a 40 km SSMF line

The fundamental limitation in the high-speed (over 2.5 Gbit/s per channel) systems is set by the total dispersion in the fiber optical transmission (FOTS) lines. Without managing the dispersion, the FOTS operation with a 10 Gbit/s transmission speed per channel is limited to the line length from 40 km to 50 km. Fig. 10 shows the eye diagrams and output signal spectra in HDWDM communication systems with 10 Gbit/s transmission speed for different frequency intervals.

From Fig. 10 it is seen that a WDM communication system with a 10 Gbit/s transmission speed per channel and a frequency interval of 50 GHz could be optimized. In the four-channel case a decrease in the frequency interval to 31.25 GHz ensures a satisfactory BER value.

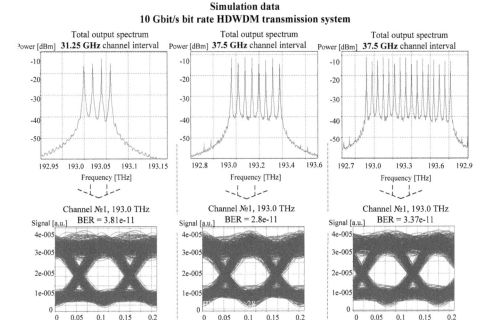

Fig. 10. 10 Gbit/s HDWDM communication system with 4/8/16 channels and 31.25/37.5/37.5 GHz frequency intervals. The output spectrum of the optical signal and eye diagrams are shown for the end of a 40 km SMF line.

With the number of channels increasing this value also increases. As a result, the optimal channel spacing in WDM systems with a 10 Gbit/s transmission speed per channel is 37.5 GHz; the possibility exists to provide a high-quality transmission of signals over 40 - 50 km (see Fig. 11).

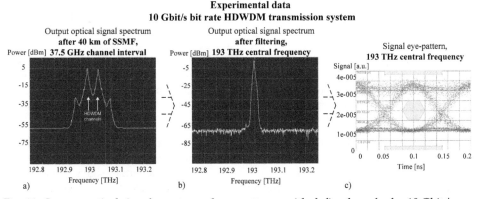

Fig. 11. Output optical signal spectra and eye-patterns with defined masks for 10 Gbit/s system: *a*) common optical spectra, *b*) signal optical spectrum after filtering, *c*) eye-pattern.

In the process of investigation it has been established that the operators of telecommunication networks when using a WDM communication system's facilities are raising its total transmission speed gradually, depending on the requested information volume, adding new channels with different transmission speeds (2.5 Gbit/s or 10 Gbit/s), different coding formats (NRZ, RZ or Duobinary) and variable frequency intervals (12.5 GHz, 25 GHz, 50 GHz, or 100 GHz); all this can result in the spectral overlapping and increased BER of the signal. We therefore have estimated the possibilities of a mixed HDWDM communication system applying different signal coding techniques (NRZ, RZ and Duo) in each channel (see Fig. 12).

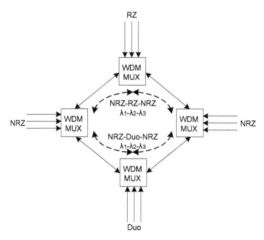

Fig. 12. A mixed scheme of the WDM communication system. The NRZ, RZ and Duobinary (Duo) intensity modulation formats are applied for the 2.5 Gbit/s and 10 Gbit/s signal transmission

The design with a transmitter unit containing three combined channels is the simplest model, in which between two NRZ signals the RZ or Duo signals are located. The NRZ signal format is chosen as a base, since it is the format preferred by the majority of telecommunication operators. Further, the operation of a mixed HDWDM communication system was subjected to scrutiny, changing the transmission speed from 2.5 Gbit/s to 10 Gbit/s per channel and the channel interval in a wide range – from 12.5 GHz to 100 GHz. Fig. 13 shows the potential of NRZ-RZ-NRZ mixed HDWDM systems (with only successful signal transmission displayed). The transmission speed used for each signal is 2.5 Gbit/s. In such a case the minimum channel interval is to be equal to or greater than 25 GHz. Only under such conditions a successful realization (i.e. with BER < 10^{-9}) is possible for a mixed HDWDM communication system with NRZ-RZ-NRZ signals.

When creating a mixed HDWDM communication system based on NRZ-RZ-NRZ formats of signals with a 10 Gbit/s transmission speed per channel it is possible to multiplex the signals with a 50 GHz frequency interval, since the quality of the output signal meets in this case the BER standard (see Fig. 14). Reducing the channel interval still further would impair the signal's characteristics. This means that the least frequency interval for the mixed NRZ-RZ-NRZ HDWDM communication system under consideration is 50 GHz.

2.5 Gbit/s bit rate NRZ-RZ-NRZ mixed HDWDM system

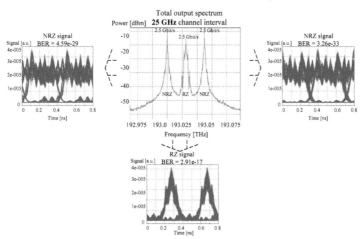

Fig. 13. A 2.5 Gbit/s mixed HDWDM communication system with NRZ-RZ-NRZ signals and a 25 GHz frequency interval. The output spectrum of the optical signal and the eye diagrams of the received electrical signal are shown

This done, the operation of a mixed HDWDM communication system with NRZ-Duobinary-NRZ signal formats was studied for a 2.5 Gbit/s transmission speed per channel. The conclusion was that it is possible to compact the signals with a 12.5 GHz frequency interval and a proper BER (< 10^{-9}); this is two times more compact than in the NRZ-RZ-NRZ case, which would provide a highly efficient use of the spectrum (see Fig. 15).

10 Gbit/s bit rate NRZ-RZ-NRZ mixed HDWDM system

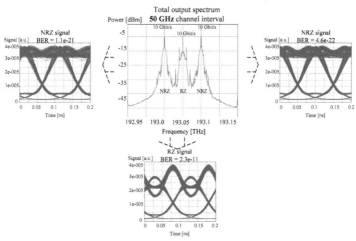

Fig. 14. A 10 Gbit/s mixed HDWDM communication system with NRZ-RZ-NRZ signals and a 50 GHz frequency interval. The output spectrum of the optical signal and the eye diagrams of the received electrical signal are shown.

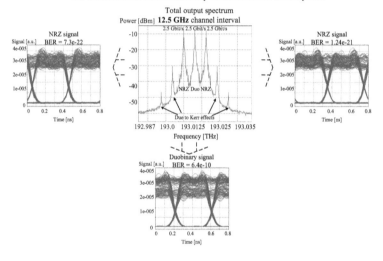

Fig. 15. A 2.5 Gbit/s mixed HDWDM communication system with NRZ-Duobinary-NRZ signals and a 12.5 GHz frequency interval. The output spectrum of the optical signal and the eye diagrams of the received electrical signal are shown.

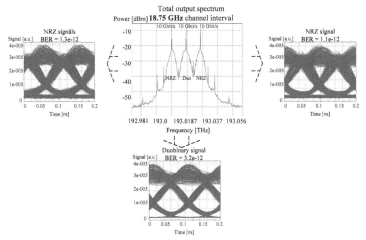

Fig. 16. A 10 Gbit/s mixed HDWDM communication system with NRZ-Duobinary-NRZ signals and an 18.75 GHz frequency interval. The output spectrum of the optical signal and the eye diagrams of the received electrical signal are shown

At the same time, the increase in the transmission speed from 2.5 Gbit/s to 10 Gbit/s in a mixed NRZ-Duo-NRZ HDWDM communication system leads to frequency interval rising up to 18.75 GHz, which, as compared with the results for a NRZ-RZ-NRZ system provides a highly efficient exploitation of the spectrum (see Fig. 16).

6. Conclusions

Our results have proved once more that HDWDM is a powerful technique for increasing the capacity of fiber optics transmission systems. It may be crucial for enabling technology of ultra-high capacity on-chip optical interconnects, as well as chip-to-chip optical interconnects in massively parallel different optical systems. It has been shown that the BER and eye-diagram technique is a good means for evaluating the system performance that allows HDWDM system to be optimized for different parameters.

In contrast to the conventional high speed approach of increasing WDM transmission capacity, we have demonstrated the minimal allowed channel spacing in HDWDM systems, and provided we are able to provide recommendations for future HDWDM solutions.

In the measurements, different optical filter FWHM values (from 0.15 nm to 0.7 nm) were used. The best results were obtained for 0.15 nm, when the eye pattern was opened wider. For evaluation of the signal quality a visual method was employed, in which the eye pattern was evaluated visually in the electric signal analyser varying the quasi-rectangular optical filter FWHM value.

At reducing the channel interval to 18.75 GHz the Kerr effects (self-phase modulation, cross-phase modulation, and four-wave mixing) degrades the 2.5 Gbit/s HDWDM system. The signal eye-pattern overlaps with the mask, which means that the signal quality does not ensure in this case the BER value of 10^{-9}. From the measurement results it follows that the 25 GHz channel interval ensures a good signal quality and that the signal eye-pattern does not overlap with the mask.

At 10 Gbit/s HDWDM transmission the channel interval should be 37.5 GHz to ensure the signal quality with the BER value of 10^{-9}, which fits well the previous simulation results.

It is established that the operators of telecommunication networks, when creating the HDWDM communication systems, raise the total transmission speed step-by-step in response to the increased request for the data volume. As a result, a mixed HDWDM system is formed, with different transmission speeds (2.5 Gbit/s or 10 Gbit/s), coding formats (NRZ, RZ or Duobinary) and frequency intervals (12.5 GHz, 25 GHz, 50 GHz, 100 GHz). Therefore, in order to ensure stabile functioning (i.e. BER < 10^{-9} for each signal) of a mixed HDWDM system the channel interval should exceed 25 GHz at a 2.5 Gbit/s transmission speed per channel. In turn, for stabile operation of a mixed 10 Gbit/s WDM system the frequency interval should be raised to 50 GHz.

The Duobinary technique for signal coding ensures a better protection of the transmitted signals against Kerr effects as compared with the RZ coding. This allows a highly compact NRZ-Duobinary-NRZ system to be formed with the 12.5 GHz frequency interval and 2.5 Gbit/s transmission speed per channel. In turn, in the case of a 10 Gbit/s transmission speed per channel it is possible to use an 18.75 GHz frequency interval.

7. References

Abbou, F., Chuah, T., Hiew, C. and Abid, A, (2008), Comparison of RZ-OOK and RZ-DPSK in Dense OTDM-WDM Systems Using Q Factor Models, *Journal of Russian Laser Research*, Vol.29, pp. 133-141.

Agrawal, G.P. (2001), *Nonlinear Fiber Optics*, 3rd edition, Academic Press, California.

Azadeh, M. (2009), *Fiber Optics Engineering*. Springer, New York.

Belai, O. V., Shapiro, D. A. and Shapiro, E. G., (2006), Optimisation of a High-Bit-Rate Optical Communication Link with a Non ideal Quasi-Rectangular Filter. *Quantum Electronics*. Vol.36(9), pp. 879-882.

Bhamber, R., Turitsyn, K., Mezentsev, V., (2007), Effect of carrier reshaping and narrow MUX-DEMUX filtering in 0.8 bit/s/Hz WDM RZ-DPSK transmission, Optical Quantum Electronics, Vol.39. pp. 687-692.

Bierlaire, M., Bolduc, D. and McFadden, D., (2007), Characteristics of generalized extreme value distributions. *Technical report.*

Binh, Le Nguyen, (2008), Photonic Signal Processing: Techniques and Applications, CRC Press, Boca Raton.

Binh, Le Nguyen, (2009), *Digital Optical Communications*, CRC Press, Boca Raton.

Bobrovs, V., Ivanovs, G., (2008), Comparison of different modulation formats and their compatibility with WDM transmission system, *Latvian Journal of Physics and Technical Sciences*, Vol.2, pp. 6-21.

Bobrovs, V., Ivanovs, Ģ., (2009) Investigation of Minimal Channel Spacing in HDWDM Systems, *Electronics and Electrical Engineering*, Vol.4(92), pp. 53-56.

Bobrovs., V., Ozolins., O., Ivanovs., G., (2010), Investigation into the potentialities of quasi-rectangular optical filters in HDWDM systems, *Latvian Journal of Physics and Technical Sciences*, Vol.1, pp. 13-25.

Cisco Systems, (2008), Cisco Visual Networking Index – Forecast and Methodology 2007-2012., *White paper*, pp. 1-15.

Chen, L., Zhang, Z. and Bao, X., (2006), Polarization dependent loss vector measurement in a system with polarization mode dispersion, *Optical Fiber Technology*, Vol.12, pp. 251-254.

Funabashi, M., Hiraiwa, K., Koizumi, S., Yamanaka, N., and Kusukawa, A., (2001), Low operating current 40 mW PM fiber coupled DFB laser modules for externally modulated 1550 nm WDM sources, *European Conference on Optical Communication*, 2(Tu.B.1.3), pp. 122-123.

Ivanovs., G., Bobrovs., V., Ozolins, O. and Porins., J., (2010), Realization of HDWDM transmission system, *International Journal of the Physical Sciences*, Vol.5(5), pp. 452-458.

Jacobsen., G., (1994) *Noise in digital optical transmission systems*, Artech House, Boston.

Kaminow, I., Tingye, L. and Willner., A., (2008), *Optical Fiber Telecommunications V A Components and Subsystems*, Elsevier Academic Press, Burlington & San Diego.

Kaminow, I., Tingye, L., and Willner., A., (2008), Optical *Fiber Telecommunications V B Systems and Networks*, Elsevier Academic Press, Burlington & San Diego.

Kashyap, R., (2010), *Fiber Bragg gratings*: Academic Press, London.

Kotz, S. and Nadarajah, S., (2000), *Extreme value distributions: theory and applications*. ICP Imperial College Press, London.

Markose, S. and Alentorn, A., (2007), *The generalized extreme value distribution, implied tail index and option pricing*, University of Essex, Colchester.

McGloin, D. and Reid, J.P., (2010), 40 Years of Optical Manipulation, *Optics and Photonics News*, Vol. 20, Iss. 3, pp. 20–26.

Ozoliņš, O., & Ivanovs, Ģ., (2009), Realization of Optimal FBG Band–Pass Filters for High Speed HDWDM, *Electronics and Electrical Engineering*, 4(92), pp. 41–44.

Ozoliņš, O., Bobrovs, V., Ivanovs, Ģ. (2011), DWDM Transmission Based on the Thin-Film Filter Technology, Latvian Journal of Physics and Technical Sciences, Vol.3, pp. 55.-65.

Pan, Z., Yu, C. and Willner, A. E., (2010), Optical performance monitoring for the next generation optical communication networks, *Optical Fiber Technology*, Vol.16 (1), pp. 20–45.

Pfennigbauer, M., and Winzer, P.J., (2006), Choice of MUX/DEMUX filters characteristics for NRZ, RZ, and CSRZ DWDM systems, *Lightwave Technology*, 24(4), pp. 1689-1696.

Rongqing, H. and O'Sullivan, M., (2009), *Fiber Optics Measurement Techniques*. Elsevier Academic Press, Burlington & San Diego.

Szodenyi, A., (2004), Optical filter type influence on transparent WDM network's size, *Hiradastechnika*, Vol.12, pp. 55–58.

Thyagarajan, K. and Ghatak, A., (2007), *Fiber Optics Essentials*, IEEE press: Willey Interscience.

Venghaus, H., (2006), *Wavelength Filters in Fibre Optics,* Springer, Berlin.

Voges, E. and Peteramann, K., (2002), *Optische Kommunikationstechnik - Handbuch für Wissenschaft un Industrie* Springer-Verlag, Berlin.

9

Design and Modeling of WDM Integrated Devices Based on Photonic Crystals

Kiazand Fasihi
Golestan University
Iran

1. Introduction

Recently, photonic crystals (PCs) have attracted great interests due to their potential ability of controlling light propagation with the existence of photonic bandgap (PBG), and the possibilities of implementing compact nanophotonic integrated circuits. Some of the most successful structures are based on planar PCs. In such structures, the optical field is confined, horizontally, by a PBG provided by the PC and, vertically, by total internal reflection due to refractive index differences. Various PC components, such as, waveguides, bends, Y splitters, directional couplers, low crosstalk intersections and all-optical switches have already been realized. These basic building blocks can be combined to realize complete circuits with various optical functions within an extremely small area. One of the most important fields for ultra-dense integrated circuits is optical communications. A key component in modern optical communications systems is a wavelength division multiplexer (WDM). This component is needed to divide and combine different wavelength channels each carrying an optical data signal. Traditionally, WDM components are realized using thin-film filters, fiber Bragg gratings (FBG), or arrayed waveguide gratings. However, such devices are not convenient for ultra-dense integration. Various concepts for realizing a WDM component utilizing the extraordinary properties of PCs have recently been proposed. These ideas include optical micro-cavities, multimode self-imaging waveguides, and superprisms, but we focus on the components which are based on the interaction of the PC micro-cavities with the waveguides.

The chapter is organized as follows: In Section 2, the hybrid waveguides are introduced and analyzed using coupled-mode theory (CMT) and the finite-difference time-domain (FDTD) methods. First, the resonance frequencies and the field distribution of the resonance modes have been analyzed, then the hybrid waveguides are introduced and analyzed using FDTD and CMT methods, and the conditions which lead to quasi-flat and Lorentzian transmission spectrum will be presented. Finally, the Fundamental approach to low cross-talk and wideband intersections design which is based on the orthogonal hybrid waveguides is presented and analyzed using CMT and FDTD methods. It will be shown that when the phase-shift of the electromagnetic waves traveling between two adjacent PC coupled cavities is approximately equal to $(k+1/2)\pi$, the best performance for the intersection can be achieved. In addition it will be shown that simultaneous crossing of ultra-short pulses is possible. In Section 3, a three-port high efficient CDF with a coupled cavity-based wavelength-selective reflector is introduced and analyzed. According to the theoretical theory using CMT in time, the performance of the proposed CDF will be investigated and the conditions which lead to 100% drop efficiency will be extracted. The performance of the designed filter will also be calculated using the 2D-FDTD method. The simulation results show that the designed CDF has a line-width of $0.78nm$ at the center wavelength $1550nm$, and also a multi-channel CDF with channel spacing around $10nm$ ($1nm$) with inter-channel crosstalk below $-30dB$ ($-15dB$) is possible. These characteristics make the proposed CDF suitable for use in WDM optical communication systems.

2. Photonic crystal hybrid waveguides: Design and modeling

As mentioned before, PCs have gained great interest due to the availability of high density integrated optical circuitry (Joannopoulos et al., 2008; Yanik et al., 2004; Niemi et al., 2006; Koshiba, 2001; Fasihi & Mohammadnejad, 2009a, 2009b; Mekis et al., 1999; Loncar et al., 2000; Martinez et al., 2003; Shin et al., 2004; Liu & Chen, 2004; Yanik et al., 2003). PC waveguides provide an efficient means of guiding light by allowing the realization of sharp optical bends (Mekis et al., 1996). By combining the PC waveguides with PC cavities, realization of compact optical filter designs is possible (Fan et al., 1999). If the cavities are brought close together, they will couple and light will propagate through evanescent wave coupling from one cavity to its neighbors. This new type of waveguide is called the coupled-cavity waveguide (CCW) or alternatively, a coupled-resonator optical waveguide (CROW) (Yariv et al., 1999) and has many interesting properties. With such waveguides, by appropriately positioning the coupled optical cavities, sharp bends are also possible. In this section, the hybrid waveguides based on combining of the CCWs and the conventional line defect waveguides proposed and modeled by using CMT and FDTD methods. PC hybrid waveguides are a key element in the construction of future integrated optical circuits. Orthogonal hybrid waveguide intersections are very good candidate for wideband and low cross-talk intersections, which are a crucial element in PC-based integrated circuits (Fasihi & Mohammadnejad, 2009a). In some applications, such as ultra-short pulse transmission and time delay lines, the transmission spectrum is designed to be quasi-flat. By increasing the confinement of the coupled cavities in hybrid waveguides, the continuous transmission band will be converted to a series of discrete bands with Lorentzian spectrum, which are useful for implementation of some optical devices, such as filters (Ding et al., 2009). In this section, both analytical and numerical approaches are used to design and modeling of the hybrid waveguides with quasi-flat and Lorentzian transmission spectrum.

2.1 Analysis of the PC cavity modes

When a local defect is created in a PC, e.g. by removing a single rod, a cavity is formed where light is confined in one or more bound states. Depending on the quality of the confinement, these states, or modes, exist only in a narrow frequency range. In general, a defect can have any shape or size; it can be made by changing the refractive index of a rod, modifying its radius, or removing a rod altogether (Villeneuve et al., 1996). The defect could also be made by changing the index or the radius of several rods. Here, we choose to modify the radius of a single rod. For ease of computation, we use a 2D-PC consisting of a hexagonal lattice of dielectric rods in air. The rods have lattice constant a, radius $r = 0.20a$, and refractive index $n_{rod} = 3.4$. This structure prohibits propagation of TM light (in-plane magnetic field) in the frequency range 0.280 to 0.452 ($2\pi c / a$). To determine the PBG regions of the PC structure, the MIT Photonic-Bands package (http://ab-initio.mit.edu/mpb) is used. In order to couple energy into the cavity, it is necessary to transfer energy through the walls of the PC. Incident light can transfer energy to the resonant mode by the evanescent field across the array of rods. The setup is shown in Fig. 1. To compute the resonant frequencies, we consider a finite-sized 13×21 PC in which a single rod has been removed. We send a wide-spectrum plane wave pulse with TM polarization at the incident angle of around 15° respect to the z-axis. On the other side of the PC, the field amplitude is monitored at a short length, marked as "Monitor". This configuration facilitates identifying of the resonance peaks of transmission spectrum, especially in some degenerate states. In this configuration the excitation has a Gaussian profile centered at $\omega = 0.37$ ($2\pi c / a$) and a width of $\Delta\omega = 0.6$ ($2\pi c / a$) which extends beyond the edges of the PBG. The resonant frequencies of the cavities are plotted as a function of the cavity radius in Fig. 2. This figure shows that as the radius of the cavity is reduced to $0.15a$, due to the increasing perturbation, a resonant cavity mode appears at the bottom of the PBG (Villeneuve et al., 1996). As the cavity radius is further reduced, since the cavity involves removing dielectric material in the PC, based on perturbation theorem, a higher frequency resonant mode is obtained (Joannopoulos et al., 2008), and eventually reaches $\omega = 0.3953$ ($2\pi c / a$) when the rod is completely removed. The corresponding electric-field (E_y) distributions of the resonant mode is shown in Fig. 3-(d) for the case $r_d = 0.1a$. We name this state as "monopole", because it has no nodal lines in the central (cavity) rod. By increasing the cavity radius to $0.25a$, three triply-degenerate dipole states appears at the top of the PBG (the field distributions are shown in Fig. 3-(a) for the case $r_d = 0.3a$).

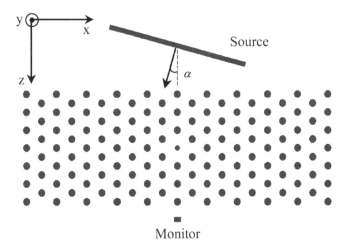

Fig. 1. The used set up for determining the resonance frequency of the cavity states in PC of hexagonal lattice.

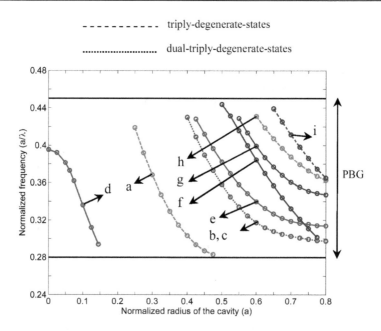

Fig. 2. The resonant frequencies for different cavities radius in the PC of hexagonal lattice.

These states named as a dipole to reflect the two lobes in their field distributions. If the radius is further increased then a sequence of higher-order modes (with more nodal planes) are pulled down into the PBG: six dual-triply-degenerate states (Fig. 3-(b), (c)), a higher-order monopole with an extra node in the radial direction (Fig. 3-(e)), a hexapole state and a second-order hexapole state (Fig. 3-(f), (g)), three triply-degenerate second order dipole states (Fig. 3-(h)), and three triply-degenerate octapole states (Fig. 3-(i)).

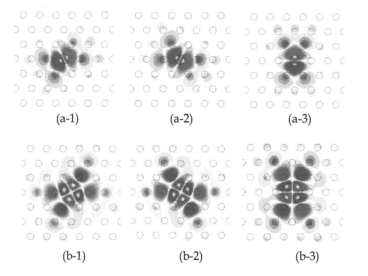

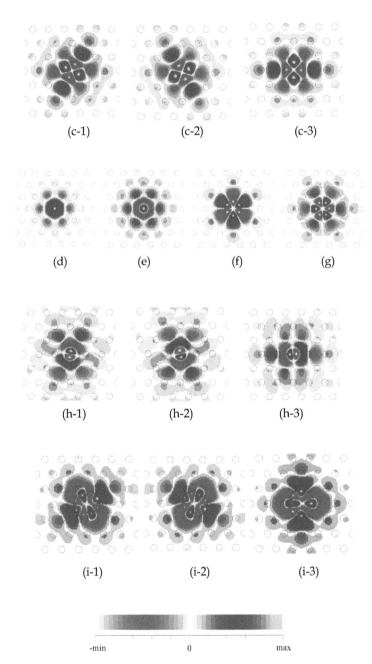

Fig. 3. The electric-field (E_y) distributions of the resonant modes for the labled points of.

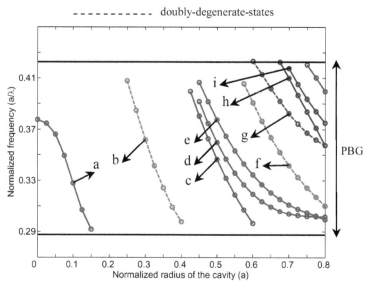

Fig. 4. The resonant frequencies for different cavities radius in the PC of square lattice.

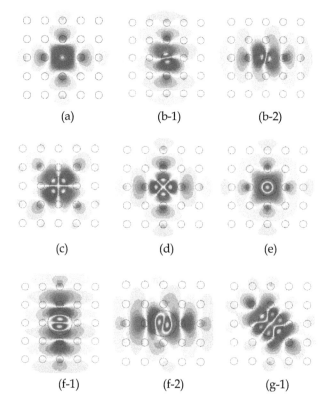

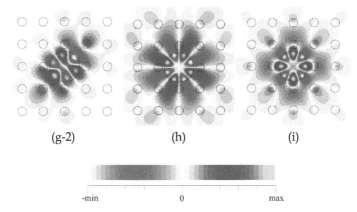

(g-2)   (h)   (i)

-min   0   max

Fig. 5. The electric-field (E_y) distributions of the resonant modes for the labled points of.

2.2 Analytical model for hybrid waveguides: CMT approximation

2.2.1 Hybrid waveguides

The PC based CCWs are formed by placing a series of high-Q optical cavities close together. In this case due to weak coupling of the cavities, light will be transferred from one cavity to its neighbors and a waveguide can be created (Yariv et al., 1999). By combining of the CCWs and the conventional line defect waveguides a new waveguide can be created, which is referred to as hybrid waveguide. Fig. 6 shows the structure of a hybrid waveguide which is implemented in a PC of square lattice. Generally, there are two types of PC lattice structures, air-hole-type and rod-type. Most theoretical studies conducted so far have investigated arrays of dielectric rods in air. The advantage of this model system is that waveguides created by removing a single line of rods are single moded. Getting light to travel around sharp bends with high transmission is then relatively straightforward, and many rod-type PC structures have been proposed. Such waveguides have been fabricated and photonic band gap guidance has been confirmed. Unfortunately, the rod in air approach does not provide sufficient vertical confinement and is difficult to implement in the optical regime.

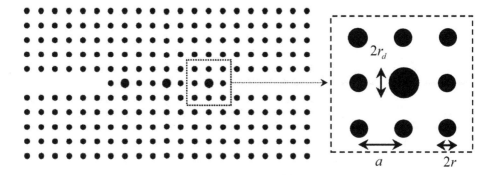

Fig. 6. Schematic of a PC hybrid waveguide of square lattice.

So, in practice, the PC slabs of dielectric rods in which the refractive index of the background material is higher than air, are used. Despite easier fabrication of PC waveguide based on air-hole-type structures than rod-type, there are limitations on frequency bandwidth of the single mode region and the group velocity (Fujisawa & Koshiba, 2006). Moreover in PC waveguides based on rod-type structure the large bandwidth and the large group velocity can be achieved, and recently such waveguides have been used for fabrication of photonic devices (Chen et al. 2005).

2.2.2 Modeling of hybrid waveguides by CMT method

Here, we consider the CCWs that are formed by periodically introducing defects along one direction in 2D-PCs. Generally, the coupling between two PC cavities depends on the leakage rate of energy amplitude into the adjacent cavity $(1/\tau)$, which defines the quality factor of the cavity, and the phase-shift between two adjacent defects (φ). In a straight CCW which contains N PC cavities $(N > 3)$, the transmission spectrum is given as (Sheng et al., 2005)

$$T_N = \left(\left[(\alpha^2 - \sin^2 \varphi) A_{N-2} - 2\alpha A_{N-3} + A_{N-4} \right]^2 \times (2\sin\varphi)^{-2} + \left[-\alpha A_{N-2} + A_{N-3} \right]^2 \right)^{-1} \quad (1)$$

where

$$\alpha = 4Q(\omega/\omega_0 - 1)\sin\varphi - \cos\varphi , Q = (\omega_0\tau)/4 . \quad (2)$$

In the above equation, ω, ω_0 and Q are the frequency of incident input, the resonant frequency and the quality factor of the PC cavities, respectively. In (1), A is a series function of $\beta = (\alpha - \cos\varphi)$ that satisfies $A_{-1} = 0$, $A_0 = 1$ and $A_m = \beta A_{m-1} - A_{m-2}$ $(m = 1,2,3,...N)$. As shown in (1), apart from ω, the transmission spectrum of a CCW depends on three parameters ω_0, Q and φ. The ω_0 and Q, can be extracted from a simple numerical simulation on a PC molecule, composed of two coupled cavities. For $N = 2$, i.e., a PC molecule, the transmission spectrum is given as

$$T_2 = \left(\left[T_{min}^{-1} - 8Q^2 \cos^2 \varphi \left(\frac{\omega}{\omega_0} - \frac{1}{4Q\tan\varphi} - 1 \right) \right]^2 + 64Q^4 \sin^2 \varphi \left(\frac{\omega}{\omega_0} - \frac{1}{4Q\tan\varphi} - 1 \right)^4 \right)^{-1} \quad (3)$$

Where

$$T_{min} = 4\left(2 + \sin^{-2}\varphi + \sin^2\varphi\right)^{-1} . \quad (4)$$

The parameter T_{min} is the minimum in transmission band of a PC molecule. The peaks of the equation (3), which are equal to unity, appear at $\omega = \omega_0$ and $\omega = \omega_0\left[1 + (2Q\tan\varphi)^{-1}\right]$. Hence, using (4) and the simulated transmission spectrum of one PC molecule, ω_0 and Q can be extracted. It must be noted that the analytical results of equation (1) can be extended to CCWs of any dimensions (Sheng et al., 2005). Now, we consider the hybrid waveguides that contain N identical cavities in 2D-PCs and generalize CMT analytical method to obtain the transmission spectrum. According to (1), it can be seen that for a given N, the transmission spectrum curve has 2N -1 number of extremums and the minimum in transmission band (T_{min}), is independent of ω_0 and Q, and we can obtain φ as a function of the radius of the coupled cavities (r_d), as follows (Fasihi & Mohammadnejad, 2009a):

- The relationship between T_{min} and r_d can be calculated by repeating a numerical simulation, such as FDTD method, for different values of r_d.

Here we consider a hybrid waveguide which contains three coupled cavities, $N = 3$ (see Fig. 6), in the 2D-PC of square lattice composed of dielectric rods in air. Now, we have chosen to name this hybrid waveguide HW3 and extend this naming to other hybrid waveguides. The rods have refractive index $n_{rod} = 3.4$ and radius $r = 0.20a$, where a is the lattice constant. By normalizing every parameter with respect to the lattice constant a, we can scale the waveguide structure to any length scale simply by scaling a. The radius of the coupled cavities are varied from $0.27a$ to $0.345a$. The grid size parameter in the FDTD simulation is set to $0.046a$ and the excitations are electromagnetic pulses with Gaussian envelope, which are applied to the input port from the left side. All the FDTD simulations below are for TM polarization. The field amplitude is monitored at suitable location at the right side of the HW3. Table 1 shows the relationship between T_{min} and r_d for the HW3 which are obtained from the FDTD simulations.

- The relationship between T_{min} and φ for the HW3 can be calculated from (1).
- Fig. 7 shows this relationship over one-half period of (1). Therefore, the relationship between φ and r_d of the HW3 can be demonstrated in Fig. 8.

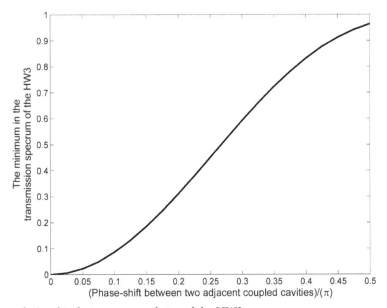

Fig. 7. The relationship between φ and T_{min} of the HW3.

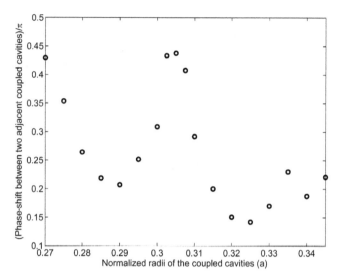

Fig. 8. The phase-shift between two adjacent cavities as a function of the cavities radius of the HW3 in a PC of square lattice (Fasihi & Mohammadnejad, 2009a).

r_d	T_{min}	r_d	T_{min}
0.270a	0.8831	0.3075a	0.8463
0.275a	0.7311	0.310a	0.5700
0.280a	0.4911	0.315a	0.3112
0.285a	0.3619	0.320a	0.1870
0.290a	0.3297	0.325a	0.1669
0.295a	0.4552	0.330a	0.2334
0.300a	0.6170	0.335a	0.3997
0.3025a	0.8894	0.340a	0.2773
0.3050a	0.8958	0.345a	0.3681

Table 1. Values of the Minimum in the HW3 Transmission Bandfor Various Radii of the Coupled Cavities

In order to compare the results of CMT and FDTD methods, we consider a HW2 under the same conditions as mentioned previously and utilize the FDTD simulation results to compute ω_0 and Q. The radius of the coupled cavities are set to $r_d = 0.32a$. The transmission spectrum of HW2 computed by the FDTD is shown in Fig. 9. According to this figure, the parameters ω_0, Q and φ are equal to 0.3428 ($2\pi c / a$), 130.3 and 0.4066π, respectively. Hence, the CMT transmission spectrum can be calculated from (3) (see Fig. 9). It is observed that the transmission spectrum calculated by CMT is in good agreement with that simulated by FDTD. As another example, we take a HW3 under the same condition as mentioned previously, with $r_d = 0.32a$ which corresponds to $\varphi = 0.1509\pi$. The transmission spectra of the above HW3 simulated by FDTD and CMT are shown in Fig. 10 for comparison. Although there is a difference in the minimum transmission spectrum between the first and second peaks, it is observed that the spectrum calculated by analytical method is nearly in good agreement with that simulated by the numerical simulation. Now, we consider a hybrid waveguide which contains three coupled cavities, in the 2D-PC of hexagonal lattice composed of dielectric rods in air. All conditions are the same as the previous structure and the radius of the coupled cavities is varied from 0 to 0.08a. The relationship between φ and r_d of the HW3 in PC of hexagonal lattice is shown in Fig. 11. Tables 2 and 3 show the transmission regions and $-3dB$ bandwidths (BW) of the proposed hybrid waveguide for different values of the coupled cavities radii in PC of square and hexagonal lattices, respectively.

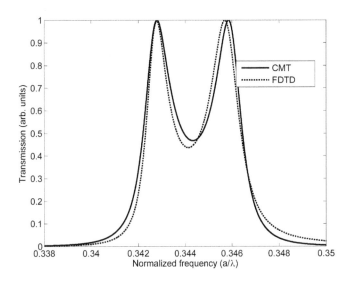

Fig. 9. The simulation results of transmission spectrum of the HW2 in PC of square lattice obtained by FDTD and CMT methods (Fasihi & Mohammadnejad, 2009a).

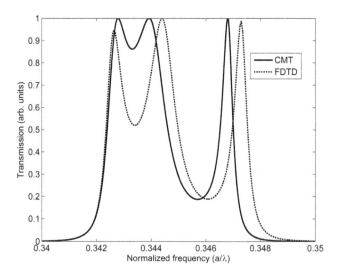

Fig. 10. The simulation results of transmission spectrum of the HW3 in PC of square lattice obtained by FDTD and CMT methods (Fasihi & Mohammadnejad, 2009a).

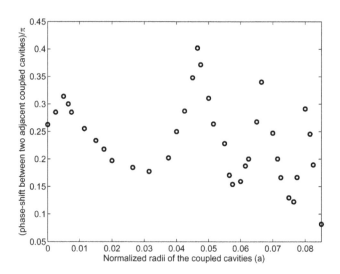

Fig. 11. The phase-shift between two adjacent cavities as a function of the cavities radius of the HW3 in a PC of hexagonal lattice.

Radius of cavities	Transmission region($2\pi c/a$)	-3dB BW (in terms of wavelength)	-3dB BW for α=0.55μm
0.27a	0.3861-0.3932	0.0468a	25.7 nm
0.28a	0.3769-0.3829	0.0415a	22.8 nm
0.29a	0.3680-0.3734	0.0393a	21.6 nm
0.30a	0.3592-0.3645	0.0404a	22.2 nm
0.3025a	0.3569-0.3624	0.0425a	23.4 nm
0.3050a	0.3545-0.3601	0.0438a	24.1 nm
0.3075a	0.3522-0.3579	0.0452a	24.9 nm
0.31a	0.3503-0.3555	0.0417a	22.9 nm
0.32a	0.3423-0.3450	0.0228a	12.5 nm
0.33a	0.3348-0.3381	0.0291a	16.0 nm
0.34a	0.3272-0.3305	0.0305a	16.8 nm

*Assuming the lattice constant α=0.55μm considering that in this case the center wavelength of transmission band is equal to 1550nm when r_d=0.3075α, the intersection BWs for different radius of the coupled cavities at working wavelength of 1550nm can be obtained and is shown in the column 4.

Table 2. Values of the Transmission Region and -3dB BW of the HW3 for Various Radiuses of the Coupled Cavities in PC of square lattice

Radius of cavities	Transmission region ($2\pi c/a$)	-3dB BW (in terms of wavelength)	-3dB BW for α=0.5937μm
0	0.3954-0.3968	0.0143a	8.4 nm
0.005a	0.3929-0.3964	0.0223a	13.2 nm
0.010a	0.3926-0.3961	0.0222a	13.1 nm
0.015a	0.3930-0.3954	0.0149a	8.8 nm
0.020a	0.3924-0.3945	0.0136a	8.0 nm
0.025a	0.3914-0.3923	0.0059a	3.5 nm
0.030a	0.3901-0.3911	0.0063a	3.7 nm
0.035a	0.3886-0.3896	0.0065a	3.8 nm
0.040a	0.3864-0.3885	0.0135a	8.0 nm
0.045a	0.3829-0.3861	0.0219a	13.0 nm
0.0465a	0.3822-0.3854	0.0220a	13.1 nm
0.0475a	0.3814-0.3846	0.0221a	13.1 nm
0.050a	0.3791-0.3823	0.0223a	13.2 nm
0.055a	0.3751-0.3778	0.0089a	5.3 nm
0.060a	0.3738-0.3744	0.0042a	2.5 nm
0.065a	0.3694-0.3720	0.0186a	11.0 nm
0.070a	0.3642-0.3653	0.0083a	4.9 nm
0.075a	0.3607-0.3615	0.0060a	3.5 nm

** Assuming the lattice constant α=0.5937μm considering that in this case the center wavelength of transmission band is equal to 1550nm when r_d=0.0475α, the intersection BWs for different radius of the coupled cavities at working wavelength of 1550nm can be obtained and is shown in the column 4.

Table 3. Values of the Transmission Region and -3dB BW of the HW3 for Various Radiuses of the Coupled Cavities in PC of hexagonal lattice

2.3 Design of hybrid waveguides with quasi-flat transmission spectrum

As discussed earlier, usually, the transmission spectrum is designed to be quasi-flat within the transmission region for various applications, such as ultra-short pulse transmission and also time delay lines. By comparing the results of Fig. 8 and Table 2, and also the results of Fig. 11 and Table 3, it can be seen that the optimum values of bandwidth is obtained when the phase-shift between two adjacent cavities are close to $\pi/2(rad)$. In this case, the transmission spectrum of the HW3 is quasi-flat. Fig. 12 shows the FDTD simulation results of the HW3 transmission spectrum (in PC of square lattice) for different radii, in which the phase-shift between the adjacent cavities is nearly $\pi/2(rad)$.

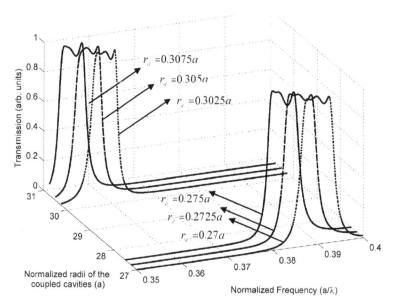

Fig. 12. The transmission behavior of the HW3 in a PC of square lattice when $\varphi \approx (k+1/2)\pi$.

2.4 Design of hybrid waveguides with Lorentzian transmission spectrum

Using (1), it can be shown that the transmission spectrum is Lorentzian if the phase-shift between the adjacent cavities is close to 0 or π. Fig. 8 shows that for different radii of the coupled cavities, the phase-shift between the adjacent cavities isn't close to zero, hence we have a continuous transmission spectrum. By increasing the confinement of the coupled cavities, the continuous transmission spectrum tends to reduce into a series of discrete modes, which are useful for some optical devices, such as the WDM filters. To do this, we place two extra rods in both ends of the CCW. We consider the structure shown in Fig. 13, and investigate the effect of increasing confinement on transmission property with locating extra rods with radius of 0.2a. All conditions are the same as the previous structure studied at section 3.2 and the radius of the coupled cavities is varied from 0 to 0.08a. Fig. 14 shows the relationship between φ and r_d of the modified HW3 with Lorentzian spectrum. The

FDTD simulation results of the modified HW3 transmission spectrum for different radii, in which the phase-shift between the adjacent cavities is nearly zero, is shown in Fig. 15.

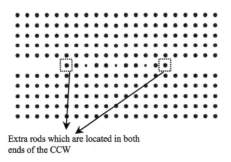

Extra rods which are located in both
ends of the CCW

Fig. 13. The modified HW3 in a PC of square lattice with Lorentzian transmission spectrum.

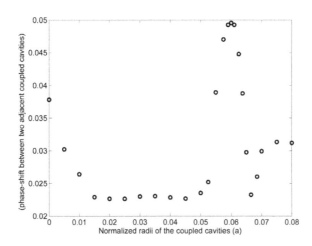

Fig. 14. The phase-shift between two adjacent cavities as a function of the cavities radius of the modified HW3 in a PC of square lattice.

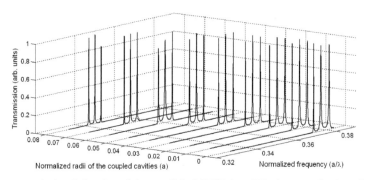

Fig. 15. The transmission behavior of the modified HW3 in a PC of square lattice when $\varphi \approx 0$.

2.5 Orthogonal hybrid waveguides: An approach to low cross-talk and wideband intersection design

In the implementation of PC-based integrated circuits, such as those used in WDM systems, it is necessary to have intersections in which crossing of ultra-short lightwave signals are possible. In another study, we show that the orthogonal hybrid waveguide intersections are very good candidate for wideband and low cross-talk intersections, which are a key element in PC-based integrated circuits (Fasihi & Mohammadnejad, 2009a). In 1998 Johnson et al. proposed a scheme to eliminate cross-talk for a waveguide intersection based on a 2D-PC of square lattice by using a single defect with doubly degenerate modes (Johnson et al., 1998). They also presented general criteria for designing such waveguide intersections based on symmetry consideration. Lan and Ishikawa presented another mechanism where the defect coupling is highly dependent on the field patterns in the defects and the alignment of the defects (i.e., the coupling angle) (Lan & Ishikawa, 2002). They asserted that their design leads to a 10 nm wide region at the central wavelength of 1310 nm with cross-talk as low as -10 to -45 dB, while in Ref. (Johnson et al., 1998) the width of the transmission band with comparable cross-talk is only 7.8 nm. In the above mentioned design, the central wavelength value of the low cross-talk transmission band is related to the air-holes radii of PC structure and therefore, adjusting the wavelength domain of transmission band is a challenge. Furthermore, Liu et al. proposed another waveguide intersection for lightwaves with no cross-talk and excellent transmission which was based on non-identical PC coupled resonator optical waveguide (CROW), without transmission band overlap (Liu et al., 2005). Li et al. proposed a different approach that utilizes a vanishing overlap of the propagation modes in the waveguides created by line defects which support dipole-like defect modes (Li et al., 2007). They claimed that in their design, over a BW of 30 nm with the central wavelength at 1300 nm, transmission efficiency above 90% with cross-talk below -30 dB can be obtained. It is obvious that in that proposal - and also in (Liu et al., 2004), simultaneous propagation of lightwaves with equal frequencies through the intersection is impossible and due to using of taper structure to solve the mode mismatch problem, total length of the intersection is increased. In our solution an approach to design of low cross-talk and wideband PC waveguide intersections based on two orthogonal hybrid waveguides in a crossbar configuration, is proposed. Without losing generality, once again we consider a 2D square lattice of infinitely long dielectric rods in the air. Fig. 16 shows the structures of an orthogonal hybrid waveguide intersection in which the rods have refractive index $n_{rod} = 3.4$ and radius $r = 0.20a$.

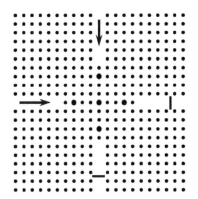

Fig. 16. Schematic structures of an orthogonal hybrid waveguide intersection.

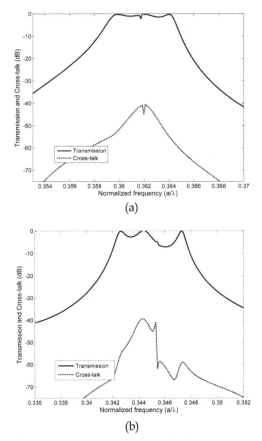

Fig. 17. The transmission and cross-talk characteristics of the orthogonal HW3 intersection when the radius of the coupled cavities are set to (a) $r_d = 0.28a$ and (b) $r_d = 0.32a$ (Fasihi & Mohammadnejad, 2009a).

Radius of cavities	-3dB BW for $a = 0.55\mu m$	Cross-talk range (dB)	
0.27a	25.7 nm	-34.35	-41.74
0.28 a	22.8 nm	-32.16	-47.60
0.29 a	21.6 nm	-33.39	-52.76
0.30 a	22.2 nm	-40.19	-53.42
0.3025 a	23.4 nm	-43.96	-55.06
0.305 a	24.1 nm	-45.35	-55.05
0.3075 a	24.9 nm	-46.66	-56.23
0.31 a	22.9 nm	-46.16	-55.58
0.32 a	12.5 nm	-39.34	-59.21
0.33 a	16.0 nm	-35.61	-60.89
0.34 a	16.8 nm	-38.79	-50.27

Table 4. Values of -3dB BW and cross-talk in orthogonal HW3 intersection for various radii of the coupled cavities

To evaluate the performance of the proposed device, the FDTD method is used for simulation, under the same conditions as mentioned previously. The excitations are electromagnetic pulses with Gaussian envelope, which are launched to the input port from the left side. The field amplitudes are monitored at suitable locations around the intersection in horizontal and perpendicular waveguides. Fig. 17-(a) and (b) shows the transmission and cross-talk characteristics of the orthogonal HW3 intersection, where the radius of the coupled cavities are set to $r_d = 0.28a$ and $r_d = 0.32a$, respectively. As can be seen from Fig. 17-(a) and (b), there exists around $0.0415a$ and $0.0228a$ regions in which the transmission is over 50%. Also, it must be noted that that the transmission properties of the proposed intersection are the same as transmission properties of the corresponding hybrid waveguide. Furthermore, by varying the radius of the coupled cavities of the hybrid waveguides, a wide frequency domain of transmission band will be obtained which proves the flexibility of the proposed design. Table II shows -3 dB BW and the cross-talk of the proposed intersection for different values of the coupled cavities radii. By comparing the results of Fig. 8 and Table 4, it can be seen that the optimum values of BW and cross-talk are obtained when $\varphi \approx (k + 1/2)\pi$. In this case, the transmission spectra of the intersection is quasi-flat (see Fig. 12).

2.5.1 Simultaneous crossing of lightwave signals and transmission of ultra-short pulses through the orthogonal hybrid waveguide intersections

In the implementation of PC-based integrated circuits, it is necessary to have intersections in which simultaneous crossing of lightwaves is possible. In the orthogonal hybrid waveguide intersections, lightwave signals can cross through the intersection simultaneously because each resonant state of the intersection will couple to modes in just one waveguide and be orthogonal to modes in the other waveguide. We consider the structure shown in Fig. 16 and verify this idea by using the FDTD technique. In this simulation, the radius of the coupled cavities of the orthogonal HW3 are chosen to be $r_d = 0.3075a$ where $a = 0.55\mu m$. During simulation, two input pulses with Gaussian envelope are applied to input ports from the top and the left sides. The monitors are placed at right and bottom output ports at suitable locations. The intensities of 500-fs pulses are adjusted to unity and 0.5, while their central wavelengths are set at 1550nm and the phase difference between them is 180°. Fig. 18 shows the transmission

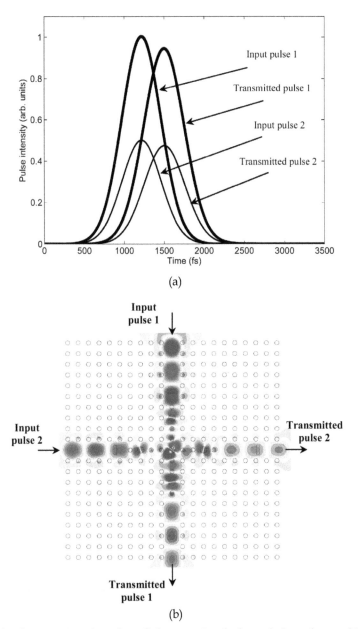

Fig. 18. The simultaneous crossing of two lightwave signals through the orthogonal HW3 intersection with $r_d = 0.3075a$ and $a = 0.55\mu m$. (a) Calculated transmission spectra (b) Calculated field distribution. The intensities of 500-fs pulses are adjusted to unity and 0.5, while their central wavelengths are set at $1550nm$ and the phase difference between them is $180°$. (Fasihi & Mohammadnejad, 2009a)

behavior of simultaneous crossing of lightwave signals through the orthogonal HW3 intersection. It can be seen that the input pulses are transmitted through the intersection with negligible interference effect. In a separate assessment, we again consider the structure shown in Fig. 16 with $r_d = 0.3075a$ where $a = 0.55\mu m$, and investigate the transmission property of the intersection for ultra-short pulses by using the FDTD method. Fig. 19 shows the transmission behavior of a 200-fs pulse whose central wavelength is 1550nm. We can see that not only the cross-talk is negligible, but also the distortion of the pulse shape is very small.

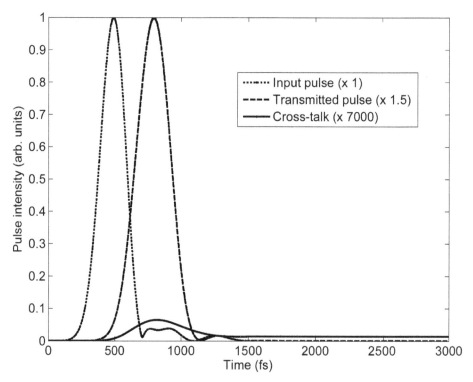

Fig. 19. The transmission behavior of a 200-fs pulse whose central wavelength is 1550nm through the orthogonal HW3 intersection with $r_d = 0.3075a$ and $a = 0.55\mu m$.

3. Highly efficient channel-drop filter with a coupled cavity-based wavelength-selective reflection feedback

The rapidly growing use of optical WDM systems, calls for ultra-compact and narrowband channel-drop filters (CDF). In a CDF, a single channel with a narrow line-width can be selected, while other passing channels remain undisturbed. The means to control the propagation of light is mainly obtained by introducing defects in PCs. Micro-cavities formed by point defects and waveguides formed by line defects in PCs. In particular, the resonant CDFs implemented in PC, which are based on the interaction of waveguides with micro-cavities, can be made ultra-compact and highly wavelength-selective (Zhang & Qiu, 2006). These devices attract strong interest due to their substantial demand in WDM optical communication systems. So far, different designs of CDFs in 2D-PCs have been proposed (Kim et al., 2007). These designs can be basically classified into two categories: surface emitting designs and in-plane designs. The surface emitting designs make use of side-coupling of a cavity to a waveguide. The input signal at resonant frequency tunnels from the waveguide into the cavity and is emitted vertically into the air (Noda et al., 2000; Song et al., 2005). The in-plane designs usually may be classified into two categories: four-port CDF designs and three-port CDF designs. The four-port CDF designs usually involve the resonant tunneling through a cavity with two degenerate modes of different symmetry, which is located between the two parallel waveguides (bus and drop). Although in this design a complete channel-drop transfer at resonant frequency is possible (i.e., 100% channel-drop efficiency), but enforcing degeneracy between the two resonant modes of different symmetry requires a complicated resonator design (Fan et al., 1998; Min et al., 2004). The operation principle of four-port CDF designs with and without mirror-terminated waveguides, have matured over the years (Zhang & Qiu, 2004). The basic concept of three-port CDF designs is based on direct resonant tunneling of input signal from bus waveguide to drop waveguide. This kind of CDF designs have simple structures and can be easily extended to design multi-channel drop filters (Kim et al., 2004), (Tekeste & Yarrison-Rice, 2006; Notomi et al., 2004). In a typical three-port CDF, the power transmission efficiency is inherently less than 50% (which corresponds the transmission in the resonant frequency), because a part of trapped signal in the cavity is reflected back to the bus waveguide when channel-drop tunneling process occurs. So far, different approaches have been proposed to solve this problem. Fan et al. proposed an approach to enhance the drop efficiency using controlled reflection to cancel the overall reflection in a full demultiplexer system. This structure is realized by coupling among an ultra low-quality factor cavity and micro-cavities with high-quality factor (Jin, 2003). Kim et al. proposed a three-port CDF with reflection feedback, in which nearly 100% drop efficiency can be theoretically achieved. In this design, the reflected back power to input port, except at the resonant frequencies, is close to 100% which leads to noise if the designed structure is incorporated in photonic integrated circuits (Kim et al., 2004). A similar design has also been proposed by Kuo et al. based on using high Q-value micro-cavities with asymmetric super-cell design (Kuo et al., 2006). This design leads to an improvement in the drop efficiency and the full-width at half-maximum (FWHM), respect to the corresponding symmetric super-cell. Another three-port CDF with a wavelength-selective reflection micro-cavity has been proposed by Ren et al. (Ren et al., 2006). In the proposed design two micro-cavities are used. One used for a resonant tunneling-based CDF, and another is used to realize wavelength-selective reflection feedback. In this section we study a three-port system which is based on two coupled cavities in both drop and reflector sections. We show that the proposed structure can provide a practical approach to attain a high efficient CDF with narrow FWHM, with no reduction in transmission efficiency parameter. Here, we consider the structure shown in Fig. 20, where the coupled cavities of the drop and the reflector sections are located at opposite sides of the bus waveguide to prevent the direct coupling between them.

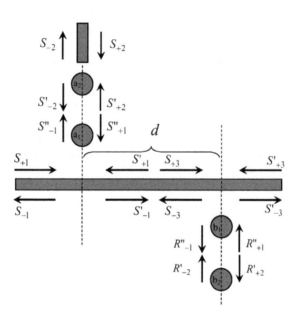

Fig. 20. The basic structure of the proposed three-port CDF with coupled cavity based wavelength selective-reflector (Fasihi & Mohammadnejad, 2009b).

The time evolution of the cavities modes, given that all of the cavities decay rates which are due to internal loss of energy be equal to τ_0, are expressed by (Fasihi & Mohammadnejad, 2009b; Haus, 1984; Manolatou et al., 1999)

$$\frac{da_1}{dt} = j\omega_{Res-a}a_1 - a_1\frac{1}{\tau_0} - a_1\frac{2}{\tau_1} - a_1\frac{1}{\tau_2} + e^{j\theta_1}\sqrt{\frac{2}{\tau_1}}S_{+1} + \sqrt{\frac{2}{\tau_2}}S''_{+1} + e^{j\theta_1}\sqrt{\frac{2}{\tau_1}}S'_{+1} \tag{5}$$

$$\frac{da_2}{dt} = j\omega_{Res-a}a_2 - a_2\frac{2}{\tau_2} - a_2\frac{1}{\tau_0} + \sqrt{\frac{2}{\tau_2}}S'_{+2} + e^{j\theta_2}\sqrt{\frac{2}{\tau_2}}S_{+2} \tag{6}$$

$$\frac{db_1}{dt} = j\omega_{Res-b}b_1 - b_1\frac{1}{\tau_0} - b_1\frac{2}{\tau_3} - b_1\frac{1}{\tau_4} + e^{j\theta_1}\sqrt{\frac{2}{\tau_3}}S_{+3} + \sqrt{\frac{2}{\tau_4}}R''_{+1} \tag{7}$$

$$\frac{db_2}{dt} = j\omega_{Res-b}b_2 - b_2\frac{1}{\tau_4} - b_2\frac{1}{\tau_0} + \sqrt{\frac{2}{\tau_4}}R'_{+2} \tag{8}$$

Here, ω_{Res-a} and ω_{Res-b} are the resonant frequencies of the coupled cavities in the drop and the reflector sections, respectively, $1/\tau_1$ and $1/\tau_3$ denote the decay rates of cavities a_1 and b_1 into the bus waveguide, respectively, $1/\tau_2$ is the decay rate of cavities a_2 into the drop waveguide and also is the decay rates of the cavity a_2 into the cavity a_1 and vice versa, and $1/\tau_4$ is the decay rates of the cavity b_2 into the cavity b_1 and vice versa. As shown in Fig.

21, the amplitudes of the electromagnetic waves (EM) incoming the drop (reflector) section from the bus waveguide, are denoted by S_{+1} (S_{+3}) and S'_{+1} (S'_{+3}). Also, the amplitudes of the EM waves outgoing the drop (reflector) section to the bus waveguide, are denoted by S_{-1} (S_{-3}) and S'_{-1} (S'_{-3}). In the case of EM waves traveling between the coupled cavities, in the drop section, the EM wave incoming the cavity a_1 (a_2) is denoted by S''_{+1} (S'_{+2}), and the EM waves outgoing the cavity a_1 (a_2) is denoted by S''_{-1} (S'_{-2}), respectively, and in the reflector section, the EM wave incoming the cavity b_1 (b_2) is denoted by R''_{+1} (R'_{+2}), and the EM wave outgoing the cavity b_1 (b_2) is denoted by R''_{-1} (R'_{-2}), respectively. The relationships among the denoted EM waves amplitudes and the cavities mode amplitudes are

$$S'_{-3} = S_{+3} - \sqrt{\frac{2}{\tau_3}} e^{-j\theta_3} b_1 \tag{9}$$

$$S_{-3} = S'_{+3} - \sqrt{\frac{2}{\tau_3}} e^{-j\theta_3} b_1 \tag{10}$$

$$S'_{-1} = S_{+1} - \sqrt{\frac{2}{\tau_1}} e^{-j\theta_1} a_1 \tag{11}$$

$$S_{-1} = S'_{+1} - \sqrt{\frac{2}{\tau_1}} e^{-j\theta_1} a_1 \tag{12}$$

$$S_{-2} = -S_{+2} + \sqrt{\frac{2}{\tau_2}} e^{-j\theta_2} a_1 \tag{13}$$

$$S''_{-1} = -S''_{+1} + \sqrt{\frac{2}{\tau_2}} a_1 \tag{14}$$

$$S'_{-2} = -S'_{+2} + \sqrt{\frac{2}{\tau_2}} a_2 \tag{15}$$

$$R''_{-1} = -R''_{+1} + \sqrt{\frac{2}{\tau_4}} b_1 \tag{16}$$

$$R'_{-2} = -R'_{+2} + \sqrt{\frac{2}{\tau_4}} b_2 \tag{17}$$

$$S_{+3} = S'_{-1} e^{-j\beta d} \tag{18}$$

$$S'_{+1} = S_{-3} e^{-j\beta d}. \tag{19}$$

In the above equations, θ_1 and θ_2 are the phases of the coupling coefficients between the bus waveguide and the cavities a_1 and b_1, respectively, θ_3 is the phase of the coupling coefficient between the drop waveguide and cavity a_2, β is the propagation constant of the bus waveguide, and d is the distance between two reference planes. The EM waves traveling between the two coupled cavities in drop and reflector sections, satisfy

$$S'_{+2} = S''_{-1}\, e^{-j\varphi} \tag{20}$$

$$S''_{+1} = S'_{-2}\, e^{-j\varphi} \tag{21}$$

$$R'_{+2} = R''_{-1}\, e^{-j\varphi} \tag{22}$$

$$R''_{+1} = R'_{-2}\, e^{-j\varphi}. \tag{23}$$

Based on Eqs. (16)-(17) and (22)-(23), when $\tau_3 = \tau_4$ we have

$$R''_{+1} = \sqrt{\frac{2}{\tau_4}} \left(\frac{b_2 - b_1\, e^{-j\varphi}}{2j\sin\varphi} \right) \tag{24}$$

$$R'_{+2} = \sqrt{\frac{2}{\tau_4}} \left(\frac{b_1 - b_2\, e^{-j\varphi}}{2j\sin\varphi} \right). \tag{25}$$

By substituting the Eqs. (24)-(25), in Eqs. (7)-(8), when EM wave is launched only from the left side into the bus waveguide (S_{+2}, $S'_{+3} = 0$), we find

$$S_{+3} = \frac{e^{-j\theta_3}\left[\gamma^2 - \dfrac{\tau_4}{\tau_0}\left(\dfrac{2\tau_4}{\tau_3} + \dfrac{\tau_4}{\tau_0} \right)\sin^2\varphi - 1 - j\gamma\sin\varphi\left(\dfrac{2\tau_4}{\tau_3} + \dfrac{2\tau_4}{\tau_0} \right) \right]}{-j\sqrt{\dfrac{2\tau_4^2}{\tau_3}}\,\sin\varphi\left[\gamma - j\left(\dfrac{\tau_4}{\tau_0} \right)\sin\varphi \right]}\, b_1 \tag{26}$$

where $\gamma = \left[(\omega - \omega_{Res-b})\tau_4 \sin\varphi - \cos\varphi \right]$. Using Eqs. (10), (18)-(19), and (26) the reflectivity and S'_{+1} can be written as

$$\frac{S_{-3}}{S_{+3}} = r = \frac{\dfrac{2\tau_4}{\tau_3}\sin\varphi\left[\gamma - j\left(\dfrac{\tau_4}{\tau_0} \right)\sin\varphi \right]}{\left[\gamma^2 - \dfrac{\tau_4}{\tau_0}\left(\dfrac{2\tau_4}{\tau_3} + \dfrac{\tau_4}{\tau_0} \right)\sin^2\varphi - 1 \right] j + \gamma\sin\varphi\left(\dfrac{2\tau_4}{\tau_3} + \dfrac{2\tau_4}{\tau_0} \right)} \tag{27}$$

$$S'_{+1} = -r\, e^{-j\rho}\left(S_{+1} - \sqrt{\frac{2}{\tau_1}}\, e^{-j\theta_1}\, a_1 \right) \tag{28}$$

where $\rho = 2\beta d$. The frequencies of the reflectivity peaks, given that $\tau_0 \gg \tau_3, \tau_4$, can be determined as

$$\omega_{\mathrm{Res}_{1,2}} = \omega_{\mathrm{Res}-b} + \frac{1}{\tau_4}\left(\frac{1}{\tan\varphi} \pm \frac{1}{\sin\varphi}\right). \tag{29}$$

From Eqs. (5), (15) and (28) the transmission spectrum of the CDF can be expressed as

$$D = \frac{S_{-2}}{S_{+1}} = \frac{\left(2/\sqrt{\tau_1\tau_2}\right)\, e^{j(\theta_1-\theta_2)}\left[(1-r\cos\rho)+j(r\sin\rho)\right]\left(j\tau_4/\tau_0\sin\varphi-\gamma\right)^{-1}}{j\left[(\omega-\omega_{\mathrm{Res}-a})+\dfrac{2r}{\tau_1}\sin\rho-\dfrac{1}{\tau_2\tan\varphi}-\dfrac{\gamma}{\alpha\tau_2\sin\varphi}\right]+\left[\dfrac{1}{\tau_0}+\dfrac{2}{\tau_1}-\dfrac{2r}{\tau_1}\cos\rho+\dfrac{(\tau_2/\tau_0+1)}{\alpha\tau_2}\right]} \tag{30}$$

where $\alpha = \gamma^2 + \sin^2\varphi\,(\tau_2/\tau_0+1)^2$. Assuming that $\varphi \approx 0$, Eq. (30) can be much simplified as

$$D\Big|_{\varphi\approx 0} = \frac{\left(2/\sqrt{\tau_1\tau_2}\right)e^{j(\theta_1-\theta_2)}\left[(1-r\cos\rho)+j(r\sin\rho)\right]}{j\left[(\omega-\omega_{\mathrm{Res}-a})+\dfrac{2r}{\tau_1}\sin\rho\right]+\left[\dfrac{2}{\tau_0}+\dfrac{2}{\tau_1}+\dfrac{1}{\tau_2}-\dfrac{2r}{\tau_1}\cos\rho\right]}. \tag{31}$$

Thus, given that $\omega_{\mathrm{Res}-a} = \omega_{\mathrm{Res}-b} = \omega_{\mathrm{Res}}$ and $\tau_0 \gg \tau_1, \tau_2$ the drop efficiency at resonant frequencies can be expressed as

$$\eta\Big|_{\omega=\omega_{\mathrm{Res}_{1,2}}} = |D|^2\Big|_{\omega=\omega_{\mathrm{Res}_{1,2}}} = \frac{8k(1-\cos\rho)}{8k^2(1-\cos\rho)+4k(1-\cos\rho)+1} \tag{32}$$

where $k = \tau_2/\tau_1$. In this case, assuming $\rho = 2\beta d = (2n+1)\pi$ for either ω_{Res_1} or ω_{Res_2}, where n is an integer, one can see that the channel drop efficiency of 100% will be obtained when $k = \tau_2/\tau_1 = 1/4$. The dependence of the maximum of drop efficiency on k parameter is shown in Fig. 21-(a). Fig. 21-(b) shows the dependence of the maximum of drop efficiency on ρ parameter. The value of the cavities quality factor has an important role in the CDF performance.

On one hand, the cavities with high quality factor are necessary for implementation of three-port CDFs with narrow FWHM, which are the key element in WDM systems. On the other hand, at resonant frequencies and given that $\varphi \approx 0$, from Eq. (27) the reflectivity can be simplified to $r = \tau_3/(\tau_3+\tau_0)$ and in order to obtain 100% reflectivity, the condition $\tau_0 \gg \tau_3$ must be satisfied. Furthermore, concerning the sensitivity of the design and fabrication tolerance, the effect of the resonant frequency difference on the transmission spectrum is considerable. Assuming $\tau_0 \gg \tau_3$, $\omega_{\mathrm{Res}-a} \neq \omega_{\mathrm{Res}-b}$, $\rho = (2n+1)\pi$ and $k = 1/4$ from Eq. (31), it can be shown that (Fasihi & Mohammadnejad, 2009b)

$$\eta_{\max}\Big|_{\omega_{\mathrm{Res}_{1,2}}} = \frac{64}{\tau_1^2(\omega_{\mathrm{Res}-b}-\omega_{\mathrm{Res}-a})^2+64} = \frac{4}{Q_1^2\left(\dfrac{\omega_{\mathrm{Res}-b}}{\omega_{\mathrm{Res}-a}}-1\right)^2+4}. \tag{33}$$

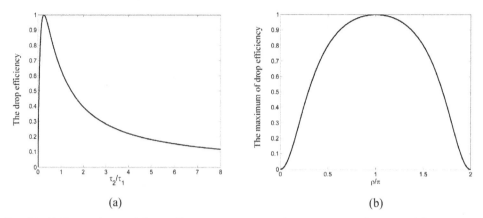

Fig. 21. (a) Dependence of drop efficiency at resonant frequencies on the ratio of decay rates τ_2 / τ_1 when $\rho = (2n+1)\pi$. (b) Dependence of the maximum of drop efficiency on ρ / π (Fasihi & Mohammadnejad, 2009b)

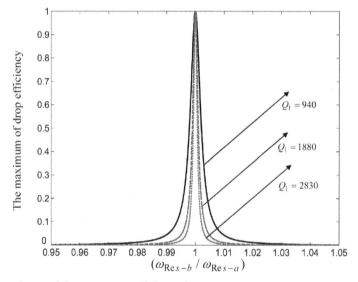

Fig. 22. Dependence of the maximum of drop efficiency on the frequency detuning factor $(\omega_{Res-b} / \omega_{Res-a})$.

This implies that by increasing the value of the quality factor, the detuning between the resonant frequencies, leads to the reduction in drop efficiency, and an advanced fabrication technology will be necessary. The drop efficiency as a function of the frequency detuning factor $(\omega_{Res-b} / \omega_{Res-a})$, is shown in Fig. 22 for modified HW1, HW2, and HW3 with $r_d = 0.04a$. Accordingly, by using appropriate structure with suitable values for the cavities quality factor, a narrowband three-port CDF with high transmission efficiency can be achieved. We investigate the validity of the proposed PC coupled cavity based CDF by employing the FDTD method with PML absorbing boundary conditions. Fig. 23 shows the structure of the three-port CDF with wavelength-selective reflection feedback, in 2D-PC of square lattice composed of dielectric rods in air. All conditions are the same as the previous structures studied at section 2. The excitations are electromagnetic pulses with Gaussian envelope, which are applied to the bus waveguide from the top side. The field amplitudes are monitored at suitable locations at the bus and the drop waveguides. Fig. 24 shows the dispersion curve of the bus and the drop line-defect waveguides versus the wave vector component k along the defect.

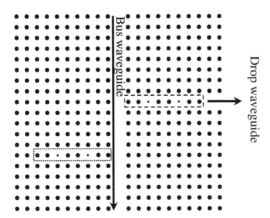

Fig. 23. The structure of three-port CDF with coupled cavity-based wavelength-selective reflection feedback, in 2D-PC of square lattice composed of dielectric rods in air. The dashed-line and the dotted-line rectangulars are the drop and the reflector sections, respectively.

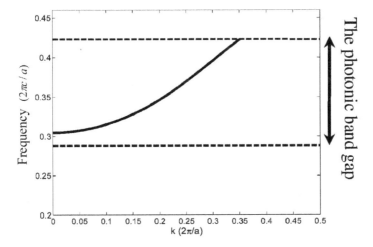

Fig. 24. Dispersion curve of the bus/drop line-defect waveguides versus the wave vector component k along the defect.

The resonant frequencies of the coupled cavities in the modified HW2 structure as a function of the coupled cavities radii are shown in Fig. 25. Given that the radii of the coupled cavities in the drop and reflector sections are set to $0.055a$, from Fig. 25, the corresponding resonant frequencies of the CDF coupled-cavities are $\omega_{\text{Res}_1} = 0.36076$ and $\omega_{\text{Res}_2} = 0.36573$ $(2\pi c/a)$. The τ_0 parameter, which is due to the internal loss of energy, is infinite in the desired 2D-PCs (Ren et al., 2006) and the total quality factors of the cavities are 1925. So, the condition $\tau_0 \gg \tau_3$ is satisfied and the perfect reflection can be realized. The condition $\tau_2/\tau_1 = 1/4$ can be easily satisfied using the coupled mode theory (Kim et al., 2004). From Fig. 25 the guided mode has wave vectors $0.2325 \times (2\pi/a)$ and $0.2428 \times (2\pi/a)$ at ω_{Res_1} and ω_{Res_2}, respectively, and when the distance between the drop and reflector sections, d, is set to $14a$, the condition $\rho(\omega_{\text{Res}_1}) = 2\beta d = (2n+1)\pi = 13\pi$ will be satisfied (in this case $\rho(\omega_{\text{Res}_2}) = 2\beta d = 13.59\pi$, which is not desired). Fig. 26-(a) shows the transmission spectra of the designed CDF calculated using the 2D-FDTD method. The simulated transmission spectrum through the drop waveguide (the dashed curve) represents that the proposed CDF has the ability of dropping a wavelength channel (at frequency ω_{Res_1}) with the dropping efficiency 0.95% and the spectral line-width 0.0014a. Assuming the lattice constant $a = 0.56\mu m$, considering that in this case the wavelength corresponds to ω_{Res_1} is equal to 1550nm when $r_d = 0.055a$, the line-width is equal to 0.78nm. Fig. 26-(b) shows the transmission spectrum of the drop waveguide in dB. In this case, it can be seen that if channel spacing, $\delta\lambda$, is chosen as $\delta\lambda > (\lambda_{\text{Res}_1} - \lambda_{\text{Res}_2})/2 \approx 10nm$, the inter channel crosstalk is reduced to below $-30\ dB$ which shows very good ability for WDM devices in practical applications.

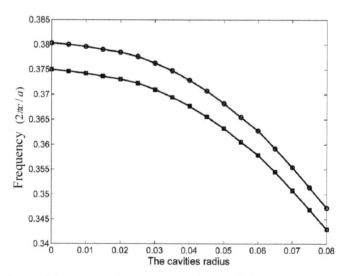

Fig. 25. Dependence of the resonant frequencies of the coupled cavities in the HW2 structure on the coupled cavities radii.

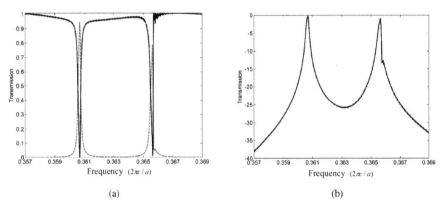

(a)   (b)

Fig. 26. Transmission spectra for the designed CDF calculated using the 2D-FDTD method. (a) The drop port (the dashed curve) and the bus port transmission spectrum (the solid curve). (b) The drop port transmission spectrum in dB (Fasihi & Mohammadnejad, 2009b).

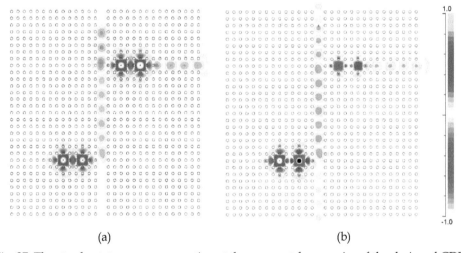

(a)   (b)

Fig. 27. The steady state wave propagation at the resonant frequencies of the designed CDF. (a) $\omega_{Res_1} = 0.36076 \ (2\pi c \ / \ a)$. (b) $\omega_{Res_2} = 0.36573 \ (2\pi c \ / \ a)$ (Fasihi & Mohammadnejad, 2009b)

The channel spacing can be reduced to $1nm$ for the $-15dB$ inter channel crosstalk. In a single cavity based CDF with reflector, the crosstalk with channel spacing of 20nm is between -18 to $-23 \ dB$ (Kuo et al., 2006). Fig. 27 shows the steady filed patterns at the resonant frequencies $\omega_{Res_1} = 0.36076$ and $\omega_{Res_2} = 0.36573 \ (2\pi c \ / \ a)$ at the bus and drop waveguides. For more optimal CDF design, the sizes of the rods between the cavities and the bus and drop waveguides, in both drop and reflector sections can be trimmed. In fact, by adjust tuning the resonant frequencies of the drop and reflector sections, further improve in CDF performance can be achieved and also the back reflection power into the input port, around the resonant frequencies, can be reduced. Even though, we don't use the additional trimming in the design. Because the add operation is the "time-reversed" process of the channel drop operation, the tunneling-based channels add and drop operation can be combined into a compact form as shown in Fig. 28. The wavelength-selective reflection section locates in the central of the structure, and it ensures full power transfer between the bus waveguide and channel-add/drop sections. The top takes the narrow-band signal out of the bus waveguide while the bottom one couples signal from the transmitters into the bus-waveguide.

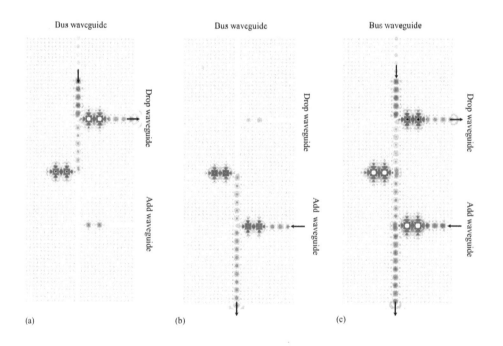

(a)   (b)   (c)

Fig. 28. The steady state wave propagation at the resonant frequency $\omega_{Res_1} = 0.36076 \, (2\pi c \, / \, a)$ of a passive channel add/drop filter designed based on proposed CDF. (a) Drop mechanism. (b) Add mechanism. (c) Add/drop mechanism.

4. Conclusions

In summary, the resonance frequencies and the field distributions of 2D-PC have been investigated. PC hybrid waveguides with quasi-flat and Lorentzian transmission spectrum were analyzed and modeled by using FDTD and CMT methods. The theoretical results derived by CMT were in good agreement with FDTD simulation results. It was shown that when the phase-shift of the electromagnetic waves traveling between two adjacent PC coupled cavities (φ), is close to $(k+1/2)\pi$, the transmission spectrum of the hybrid waveguide is quasi-flat. A modified HW3 with extra rods in both ends of the CCW and Lorentzian transmission spectrum was proposed, which can be used in implementation of WDM filters. It was shown that in this case φ is close to zero. Transmission of ultra-short pulses through the hybrid waveguide was also investigated. A low cross-talk and wideband PC waveguide intersection design based on two orthogonal hybrid waveguides in a crossbar configuration was proposed. Also, it has been shown that when $\varphi \approx (k+1/2)\pi$, optimum performance results for the intersection can be achieved. In addition, it has been clearly proved that simultaneous crossing of ultra-short pulses through the intersection is possible with negligible interference. The transmission of a 200-fs pulse at 1550 nm was simulated by using the FDTD method, and the transmitted pulse showed negligible cross-talk and very little distortion. A three-port high efficient CDF with a coupled cavity-based wavelength-selective reflector, which can be used in WDM optical communication systems, was proposed. The CMT was employed to drive the necessary conditions for achieving 100% drop efficiency. The FDTD simulation results of proposed CDF which was implemented in 2D-PC, showed that the analysis was valid. The simulation results show that the designed CDF has a line-width of $0.78nm$ at the center wavelength $1550nm$, and also a multi-channel CDF with channel spacing around $10nm$ ($1nm$) with inter-channel crosstalk below $-30dB$ ($-15dB$) is possible. These characteristics make the proposed CDF suitable for use in WDM optical communication systems.

5. References

Chen, C. C. Chen, C. Y. Wang, W. K. Huang, F. H. Lin, C. K. Chiu, W. Y. and Chan, Y. J. (2005). Photonic Crystal Directional Couplers Formed by InAlGaAs Nano-Rods. Opt. Express, 13, pp. 38–43.

Ding, W. Chen L. and Liu, S. (2004). Localization properties and the effects on multi-mode switching in discrete mode CCWs. Optics Communications., 248, pp. 479–484.

Fan, S. Villeneuve, P. R. Joannopoulos, J. D. Khan, M. J. Manolatou, C. and Haus, H. A. (1999). Theoretical analysis of channel drop tunneling processes. Phys. Rev., B (59), pp. 15882–15892.

Fasihi, K.and Mohammadnejad, S. (2009a). Orthogonal Hybrid Waveguides: an Approach to Low Cross-talk and Wideband Photonic Crystal Intersections Design. IEEE J. Lightw. Technol., 27, pp. 799–805.

Fasihi, K. and Mohammadnejad, S. (2009b). Highly efficient channel-drop filter with a coupled cavity-based wavelength-selective reflection feedback. Opt. Express., 173, pp. 8983–8997.

Fan, S. Villeneuve, P. R. Joannopoulos, J. D. and Haus, H. A. (1998). Channel drop filters in photonic crystals. Opt. Express. 3, 4-11.

Fujisawa, T. and Koshiba, M. (2006). Finite-Element Modeling of Nonlinear Mach-Zehnder Interferometers Based on Photonic- Crystal Waveguides for All-optical Signal Processing. IEEE J. Lightw. Technol., 24, pp. 617–623.

Haus, H. A. (1984). Waves and Field in Optoelectronics (Prentice-Hall,).

Jin, C. Fan, S. Han, S. and Zhang, D. (2003). Reflectionless multichannel wavelength demultiplexer in a transmission resonator configuration. J. Quantum Electron. 39, 160-165.

Joannopoulos, J. D. Johnson, S. G. Winn, J. N. and Meade, R. D. (2008). *Photonic Crystal: Molding the Flow of Light.* Princeton, Princeton Univ. Press.

Johnson, S. J. Manolatou, C. Fan, S. Villeneuve, P. R. Joannopoulos, J. D. and Haus, H. A. (1998). Elimination of Crosstalk in Waveguide Intersections," *Opt. Lett.,* vol. 23, pp. 1855-1857.

Kim, S. park, I. Lim, H. and Kee, C. (2004). Highly efficient photonic crystal-based multi-channel drop filters of three-port system with reflection feedback. Opt. Express. 12, 5518-5525.

Koshiba, M. (2001). Wavelength Division Multiplexing and Demultiplexing With Photonic Crystal Waveguide Coupler. IEEE J. Lightw. Technol., 19, pp. 1970–1975.

Kuo, C. W. Chang, C. F. Chen, M. H. Chen, S. Y. and Wu, Y. D. (2006). A new approach of planar multi-channel wavelength division multiplexing system using asymmetric super-cell photonic crystal structures. Opt. Express. 15, 198-206.

Loncar, M. Nedeljkovic, D. Doll, T. Vuckovic, J. Scherer, A. and Pearsall, T. P. (2000). Waveguiding in planar photonic crystals. Appl. Phys. Lett., 77, pp. 1937–1939.

Lan, S. and Ishikawa, H. (2002). broadband waveguide intersections with low cross talk in photonic crystal circuits. *Opt. Lett.,* vol. 27, pp. 1567–1569.

Li, Z. Chen, H. Chen, J. Yang, F. Zheng H. and Feng, S. (2007). A proposal for low cross-talk square-lattice photonic crystal waveguide intersection utilizing the symmetry of waveguide modes. *Optics Communications.,* vol. 273, pp. 89–93.

Liu, T. Fallahi, M. Mansuripour, M. Zakharian, A. R. and Moloney, V. (2005). Intersection of nonidentical optical waveguides based on photonic crystals. *Opt. Lett.,* vol. 30, pp. 2409-2411.

Liu, C. Y. and Chen, L. W. (2004). Tunable photonic-crystal waveguide Mach–Zehnder interferometer achieved by nematic liquid-crystal phase modulation. Opt. Express., 12(12) pp. 2616-2624.

Martinez, A. Griol, A. Sanchis, P. and Marti, J. (2003). Mach–Zehnder interferometer employing coupled-resonator optical waveguides. Opt. Lett., 28, no. 6, pp. 405–407.

Manolatou, C. Khan, M. J. Fan, S. Villeneuve, P. R. Haus, H. A. and Joannopoulos, J. D. (1999). Coupling of modes analysis of resonant channel add-drop filters. IEEE J. Quantum Electron. 35, 1322 -1333.

Mekis, A. Meier, M. Dodabalapur, A. Slusher, R. E. and Joannopoulos, J. D. (1999). Lasing mechanism in two-dimensional photonic crystal lasers. Appl. Phys. A: Materials Science & Processing, 69, pp. 111–114.

Mekis, A. Chen, J. C. Kurland, I. Fan, S. Villeneuve, P. R. and Joannopoulos, J. D. (1996). High transmission through sharp bendsin photonic crystal waveguides," Phys. Rev. Lett., 77, pp. 3787–3790.

Min, B. K. Kim, J. E. and Park, H. Y. (2004). Channel drop filters using resonant tunneling processes in two dimensional triangular lattice photonic crystal slabs. Opt. Commun. 237, 59-63.

Niemi, T. H. Frandsen, L. Hede, K. K. Harpøth, A. Borel, P. I. and Kristensen, M. (2006). Wavelength-Division Demultiplexing Using Photonic Crystal Waveguides. IEEE Photon. Technol. Lett., 18, pp. 226–228.

Noda, S. Chutinan, A. and Imada, M. (2000). Trapping and emission of photons by a single defect in a photonic bandgap structure. Nature. 407, 608-610.

Notomi, M. Shinya, A. Mitsugi, S. Kuramochi, E. and Ryu, H. (2004). "Waveguides, resonators and their coupled elements in photonic crystal slabs. Opt. Express. 12, 1551-1561.

Ren, H. Jiang, C. Hu, W. Gao, M. and Wang, J. (2006). Photonic crystal channel drop filter with a wavelength-selective reflection micro-cavity. Opt. Express. 14, 2446-2458.

Sheng, L. X. Wen C. X. and Sheng, L. (2005). Analysis and engineering of coupled cavity waveguides based on coupled-mode theory. Chin. Phys. Soc. and IOP Publishing Ltd., 14, pp. 2033–2040.

Shin, M. H. Kim, W. J. Kuang, W. Cao, J. R. Yukawa, H. Choi, S. J. O'Brien, J. D. Dapkus, P. D. and Marshall, W. K. (2004). Two-dimensional photonic crystal Mach–Zehnder interferometers. Appl. Phys. Lett., 84, no. 4, pp. 460–462.

Song, B. Asano, T. Akahane, Y.and Noda, S. (2005). Role of interfaces in hetero photonic crystals for manipulation of photons. Phys. Rev. B 71, 195101-19105.

Tekeste M. Y. and Yarrison-Rice, J. M. (2006). High efficiency photonic crystal based wavelength demultiplexer. Opt. Express. 14, 7931-7942.

Villeneuve, P. R. Fan, S. and Joannopoulos, J. D. (1996). Microcavities in photonic crystals: Mode symmetry, tunability, and coupling efficiency. Phys. Rev., vol. B 54, pp. 7837-7842.

Yanik, M. F. Altug, H. Vuckovic, J. and Fan, S. (2004). Submicometer All-Optical Digital Memory and Integration of Nanoscale Photonic Devices without Isolator. IEEE J. Lightw. Technol., 22, pp. 2316–2322.

Yanik, M. F. Fan, S. Soljaˇciˊc, M. and Joannopoulos, J. D. (2003). All-optical transistor action with bistable switching in a photonic crystal crosswaveguide geometry. Opt. Lett., 28(24), pp. 2506–2508.

Yariv, A. Xu, Y. Lee, R. and Scherer, A. (1999). "Coupled-resonator optical waveguide: a proposal and analysis," Opt. Lett., (24), pp. 711–713.

Zhang Z. and Qiu, M. (2004). Coupled-mode analysis of a resonant channel drop filter using waveguides with mirror boundaries. J. Opt. Soc. Am. B 23, 104-113.

Zhang Z. and Qiu, M. (2005). Compact in-plane channel drop filter design using a single cavity with two degenerate modes in 2D photonic crystal slabs. Opt. Express. 13, 2596-2604.

Part 4

Optical Communications Systems: Network Traffic

10

Traffic Engineering

Mahesh Kumar Porwal

Shrinathji Institute of Technology & Engineering, Nathdwara (Rajasthan),
India

1. Introduction

Multi Protocol Label Switching (MPLS) is today mostly used for traffic engineering therefore we start by describing what traffic engineering is and why traffic engineering is needed.

Traffic engineering and fast reroute are the two major applications of constraint based routing Traffic engineering is the process of controlling how traffic flows through a service provider's network so as to optimize resource utilization and network performance[1]. Traffic engineering is needed in the Internet mainly because the shortest path is used in current intra- domain routing protocols (e.g., OSPF, IS-IS) to forward traffic. The shortest path routing may give rise to two problems.

First, the shortest paths from different sources overlap at some links, resulting in congestion at those links.

Second, at some time, the traffic volume from a source to a destination could exceed the capacity of the shortest path, while a longer path between these two nodes remains under-utilized. The reason why conventional IP routing cannot provide traffic engineering is that it does not take into account the available bandwidth on individual links. For the purpose of traffic engineering, constraint based routing is used to route traffic trunk[2], which is defined as a collection of individual transmission control protocol (TCP), or user datagram protocol (UDP) flows, called "microflows" that share two common properties.

The **first** property is that all microflows are forwarded along the same common path.

The **second** property is that they all share the same class of service. By routing at the granularity of traffic trunks, traffic trunks have better scaling properties than routing at the granularity of individual microflows with respect to the amount of forwarding state and the volume of control traffic.

In a sense, IP networks manage themselves. A host using the Transmission Control Protocol (TCP) adjusts its sending rate according to the available bandwidth on the path to the receiver. If the network topology should change, routers react to changes and calculate new paths to the destination. This has made the TCP/IP [3] Internet a robust communication network. But robustness does not implicate that the network runs efficiently. The interior gateway protocols used today like OSPF and ISIS compute the shortest way to the destination and routers forward traffic according to the routing tables build from those calculations. This means that traffic from different sources passing through a router with the same destination will be aggregated and sent through the same path. Therefore a link may be congested despite the presence of under-utilized link in the network. And delay sensitive traffic like voice-over-IP calls may travel over a path with high propagation delay because this is the shortest path while a low latency path is available.

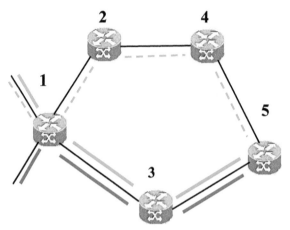

Fig. 1. Traffic Engineering

As illustrated in the above figure 1 the shortest path from router 1 to 5 is the path (1-3-5). All traffic passing through router 1 with destination router 5 (or another router with router 5 in the shortest path) will travel through this shortest path if the shortest path algorithm is used for forwarding in this network. Although there is an alternative path (1-2-4-5) available that could be used to distribute traffic more evenly in the network.

Traffic engineering is the process of controlling how traffic flows through a network to optimize resource utilization and network performance [4]. Traffic engineering is basically concerned with two problems that occur from routing protocols that only use the shortest path as constraint when they construct a routing table.

The shortest paths from different sources overlap at some links, causing congestion on those links. The traffic from a source to a destination exceeds the capacity of the shortest path, while a longer path between these two routers is under-utilized.

MPLS can be used as a traffic engineering tool to direct traffic in a network in a more efficient way then original IP shortest path routing. MPLS can be used to control which paths traffic travels through the network and therefore a more efficient use of the network resources can be achieved. Paths in the network can be reserved for traffic that is sensitive, and links and router that is more secure and not known to fail can be used for this kind of traffic.

2. Traffic engineering's role in next-generation networks

Traditional service provider networks provided Layer 2 point-to-point virtual circuits with contractually predefined bandwidth. Regardless of the technology used to implement the service (X.25, Frame Relay or ATM), the traffic engineering (optimal distribution of load across all available network links) was inherent in the process.

In most cases, the calculation of the optimum routing of virtual circuits was done off-line by a network management platform; advanced networks (offering Frame Relay or ATM switched virtual circuits) also offered real-time on-demand establishment of virtual circuits. However, the process was always the same:

- The free network capacity was examined.
- The end-to-end hop-by-hop path throughout the network that satisfied the contractual requirements (and, if needed, met other criteria) was computed.
- A virtual circuit was established along the computed path.

Internet and most IP-based services, including IP-based virtual private networks (VPNs) implemented with MPLS VPN, IPsec or Layer 2 transport protocol (L2TP), follow a completely different service model:

- The traffic contract specifies ingress and egress bandwidth for each site, not site-to-site traffic requirements.
- Every IP packet is routed through the network independently, and every router in the path makes independent next-hop decisions.
- Once merged, all packets toward the same destination take the same path (whereas multiple virtual circuits toward the same site could traverse different links).

Simplified to the extreme, the two paradigms could be expressed as follows:

- Layer 2 switched networks assume that the bandwidth is expensive and try to optimize its usage, resulting in complex circuit setup mechanisms and expensive switching methods.
- IP networks assume that the bandwidth is "free" and focus on low-cost, high-speed switching of a high volume of traffic.

The significant difference between the cost-per-switched-megabit of Layer 2 network (for example, ATM) and routed (IP) network has forced nearly all service providers to build next-generation networks exclusively on IP. Even in modern fiber-optics networks, however, bandwidth is not totally free, and there are always scenarios where you could use free resources of an underutilized link to ease the pressure on an overloaded path. Effectively, you would need traffic engineering capabilities in routed IP networks, but they are simply not available in the traditional hop-by-hop, destination-only routing model that most IP networks use.

Various approaches (including creative designs, as well as new technologies) have been tried to bring the traffic engineering capabilities to IP-based networks. We can group them roughly into these categories:

- The network core uses Layer 2 switched technology (ATM or Frame Relay) that has inherent traffic engineering capabilities. Virtual circuits are then established between edge routers as needed.
- IP routing tricks are used to modify the operation of IP routing protocols, resulting in adjustments to the path the packets are taking through the network.
- Deployment of IP-based virtual circuit technologies, including IP-over-IP tunnels and MPLS traffic engineering.

The Layer 2 network core design was used extensively when the service providers were introducing IP as an additional service into their WAN networks. Many large service providers have already dropped this approach because it does not result in the cost reduction or increase in switching speed that pure IP-based networks bring

3. Traffic engineering objectives

Traffic Engineering (TE) is concerned with performance optimization of operational networks. More formally speaking, the key traffic engineering objectives are:

1. **Minimizing congestion:** Congestion occurs either when network resources are insufficient or inadequate to accommodate offered load or if traffic streams are inefficiently mapped onto available resources; causing subsets of network resources to become over-utilized while others remain underutilized [5].
2. **Reliable network operations:** Adequate capacity for service restoration must be available keeping in mind multiple failure scenarios, and at the same time, there must be mechanisms to efficiently and speedily reroute traffic through the redundant capacity. On recovering from the faults, re-optimization may be necessary to include the restored capacity.
3. **Quality of Service requirements:** In a multi-class service environment, where traffic streams with different service requirements contend with each other, the role of traffic engineering becomes more decisive. In such scenarios, traffic engineering has to provision resources selectively for various classes of streams, judiciously sharing the network resources, giving preferential treatment to some service classes.
4. **Traffic oriented:** Traffic oriented performance objectives include the aspects that enhance the QoS of traffic streams. In a single class, best effort Internet service model, the key traffic oriented performance objectives include: minimization of packet loss, minimization of delay, maximization of throughput, and enforcement of service level agreements. Under a single class best effort Internet service model, minimization of packet loss is one of the most important traffic oriented performance objectives. Statistically bounded traffic oriented performance objectives (such as peak to peak packet delay variation, loss ratio, and maximum packet transfer delay) might become useful in the forthcoming differentiated services Internet.
5. **Resource oriented:** Resource oriented performance objectives include the aspects pertaining to the optimization of resource utilization. Efficient management of network resources is the vehicle for the attainment of resource oriented performance objectives. In particular, it is generally desirable to ensure that subsets of network resources do not become over utilized and congested while other subsets along alternate feasible paths remain underutilized. Bandwidth is a crucial resource in contemporary networks. Therefore, a central function of Traffic Engineering is to efficiently manage bandwidth resources.

4. Components of traffic engineering

One of the strategies for TE using MPLS involves four functional components [6]:

1. Information distribution
2. Path selection

3. Signaling and path set-up
4. Packet forwarding

Now, discussing each of the components in detail:

1. **Information Distribution:** Traffic engineering requires detailed knowledge about the network topology as well as dynamic information about network loading. This can be implemented by using simple extensions to IGP so that link attributes (such as maximum link bandwidth, current bandwidth usage, current bandwidth reservation) are included as part of routers link-state advertisements. The standard flooding algorithm used by link-state IGP ensures that link attributes are distributed to all routers in ISPs routing domain. Each LSR maintains network link attributes and topology information in a specialized TE database (TED), which is used exclusively for calculating explicit paths for placement of LSPs on physical topology.
2. **Path Selection:** On the basis of the network topology and link attributes in the TED and some administrative attributes obtained from user configuration, each ingress LSR calculates the explicit paths for its LSPs, which may be strict or loose. A strict explicit route is one in which the ingress LSR specifies all the LSRs in the LSP, while only some LSRs are specified in a loose explicit path. LSP calculations may also be done offline for optimal utilization of network resources.
3. **Signaling and Path-Setup:** The path calculated by the path selection component is not known to be workable, until LSP is actually established by the signaling component, because it is calculated on the basis of information present in TED, which may not be up-to-date. The signaling component is responsible for establishing LSP state and label binding and distribution in the path set-up process.
4. **Packet-Forwarding:** Once the path is set-up, packet forwarding process begins at the Label Switch Router (LSR) and is based on the concept of label switching.

5. MPLS and traffic engineering

MPLS is strategically significant for Traffic Engineering because it can potentially provide most of the functionality available from the overlay model, in an integrated manner, and at a lower cost than the currently competing alternatives. Equally importantly, MPLS offers the possibility to automate aspects of the Traffic Engineering function.

The concept of MPLS traffic trunks is used, according to Li and Rekhter [7], a traffic trunk is an aggregation of traffic flows of the same class which are placed inside a Label Switched Path. Essentially, a traffic trunk is an abstract representation of traffic to which specific characteristics can be associated. It is useful to view traffic trunks as objects that can be routed; that is, the path through which a traffic trunk traverses can be changed. In this respect, traffic trunks are similar to virtual circuits in ATM and Frame Relay networks. It is important, however, to emphasize that there is a fundamental distinction between a traffic trunk and the path, and indeed the LSP, through which it traverses. An LSP is a specification of the label switched path through which the traffic traverses. In practice, the terms LSP and traffic trunk are often used synonymously.

The attractiveness of MPLS for Traffic Engineering can be attributed to the following factors:

1. Explicit label switched paths which are not constrained by the destination based forwarding paradigm can be easily created through manual administrative action or through automated action by the underlying protocols.
2. LSPs can potentially be efficiently maintained.
3. Traffic trunks can be instantiated and mapped onto LSPs.
4. A set of attributes can be associated with traffic trunks which modulate their behavioral characteristics.
5. A set of attributes can be associated with resources which constrain the placement of LSPs and traffic trunks across them.
6. MPLS allows for both traffic aggregation and disaggregating whereas classical destination only based IP forwarding permits only aggregation.
7. It is relatively easy to integrate a "constraint-based routing" framework with MPLS.
8. A good implementation of MPLS can offer significantly lower overhead than competing alternatives for Traffic Engineering.

6. The MPLS domain

In [3] the MPLS domain is described as "a contiguous set of nodes which operate using MPLS routing and forwarding". This domain is typically managed and controlled by one administration. The MPLS domain concept is therefore similar to the notion of an AS (autonomous system), as the term is used in conventional IP routing i.e. a set of related routers that are usually under one administrative and management control.

The MPLS domain can be divided into MPLS core and MPLS edge. The core consists of nodes neighboring only to MPLS capable nodes, while the edge consists of nodes neighboring both MPLS capable and incapable nodes. The nodes in the MPLS domain are often called LSRs (Label Switch Routers). The nodes in the core are called transit LSRs and the nodes in the MPLS edge are called LERs (Label Edge Routers). If a LER is the first node in the path for a packet traveling through the MPLS domain this node is called the ingress LER, if it is the last node in a path it's called the egress LER. Note that these terms are applied according to the direction of a flow in the network, one node can therefore be both ingress and egress LER depending on which flow is considered. The terms upstream and downstream routers are also often used to indicate in which order the routers are traversed. If a LSR is upstream from another LSR, traffic is passed through that LSR before the other (downstream). A schematic view of the MPLS domain is illustrated in figure 2.

7. MPLS traffic engineering essentials

Multi-Protocol Label Switching (MPLS) is the end result of the efforts to integrate Layer 3 switching, better known as routing, with Layer 2 WAN backbones, primarily ATM. Even though the IP+ATM paradigm is mostly gone today because of the drastic shift to IP-only networks in the last few years, MPLS retains a number of useful features from Layer 2 technologies. One of the most notable is the ability to send packets across the network through a virtual circuit called Label Switched Path, or LSP, in MPLS terminology.

While the Layer 2 virtual circuits are almost always bidirectional (although the traffic contracts in each direction can be different), the LSPs are always unidirectional. If you need bidirectional connectivity between a pair of routers, you have to establish two LSPs.

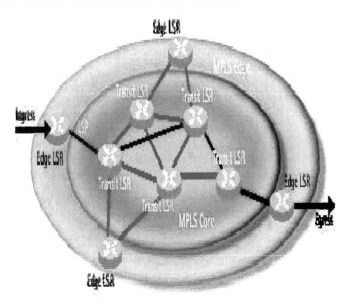

Fig. 2. The MPLS domain

The LSPs in MPLS networks are usually established based on the contents of IP routing tables in core routers. However, there is nothing that would prevent LSPs being established and used through other means, provided that:

1. All the routers along the path agree on a common signaling protocol.
2. The router where the LSP starts (head-end router) and the router where the LSP ends (tail-end router) agree on what's traveling across the LSP.

The other routers along the LSP do not inspect the packets traversing the LSP and are thus oblivious to their content; they just need to understand the signaling protocol that is used to establish the LSP.

With the necessary infrastructure in place, it was only a matter of time before someone would get the idea to use LSPs to implement MPLS-based traffic engineering. The MPLS traffic engineering technology has evolved and matured significantly since then, but the concepts have not changed much since its introduction:

1. The network operator configures an MPLS traffic engineering path on the head-end router. (The configuration mechanism involves a tunnel interface that represents the unidirectional MPLS TE LSP.)
2. The head-end router computes the best hop-by-hop path across the network, based on resource availability advertised by other routers. Extensions to link-state routing protocols (OSPF or IS-IS) are used to advertise resource availability.

NOTE: The first MPLS TE implementations supported only static hop-by-hop definitions. These can still be used in situations where you need a very tight hop-by-hop control over the path the MPLS TE LSP will take or in networks using a routing protocol that does not have MPLS TE extensions.

1. The head-end router requests LSP establishment using a dedicated signaling protocol. As is often the case, two protocols were designed to provide the same functionality as RSVP-TE (RSVP extensions for traffic engineering) and CR-LDP (constraint-based routing using label distribution protocol).
2. The routers along the path accept (or reject) the MPLS TE LSP establishment request and set up the necessary internal MPLS switching infrastructure.
3. When all the routers in the path accept the LSP signaling request, the MPLS TE LSP is operational.
4. The head-end router can use MPLS TE LSP to handle special data (initial implementations only supported static routing into MPLS traffic engineering tunnels) or seamlessly integrate the new path into the link-state routing protocol.

The tight integration of MPLS traffic engineering with the IP routing protocols provides an important advantage over the traditional Layer 2 WAN networks. In the Layer 2 backbones, the operator had to establish all the virtual circuits across the backbone (using a network management platform or by configuring switched virtual circuits on edge devices), whereas the MPLS TE can automatically augment and enhance the mesh of LSPs already established based on network topology discovered by IP routing protocols. You can thus use MPLS traffic engineering as a short-term measure to relieve the temporary network congestion or as a network core optimization tool without involving the edge routers.

In recent years, MPLS traffic engineering technology (and its implementation) has grown well beyond features offered by traditional WAN networks. For example:

1. **Fast reroute** provides temporary bypass of network failure (be it link or node failure) comparable to SONET/SDH reroute capabilities.
2. **Re-optimization** allows the head-end routers to utilize resources that became available after the LSP was established.
3. **Make-before-break** signaling enables the head-end router to provision the optimized LSP before tearing down the already established LSP.
4. **Automatic bandwidth adjustments** measure the actual traffic sent across an MPLS TE LSP and adjust its reservations to match the actual usage.

8. Requirements for traffic engineering model

A TE process model must follow a set of actions to optimize the network performance. This model has the following components:

Measurement: Measurement is an important component of the TE function. The network performance can only be determined through measurement. Traffic measurement is an essential tool to guide the network administrator of large IP networks in detecting and diagnosing performance problems, and evaluating potential control actions. The data measurement is analyzed and a decision based on the analysis is taken for network performance optimization. Measurement is needed to determine the quality of services and to evaluate TE policies.

Modelling, Analysis, and Simulation: Modelling and analysis are important aspects for TE. A network model is an abstract representation of the network that captures the network features, attributes and characteristics (e.g. link and nodal attributes). A network model can facilitate analysis or simulation, and thus can be useful to predict the network performance.

Network modelling can be classified as structural or behavioural module. Structural modules focus on the organization of the network and its components. Behavioral modules focus on the dynamics of the networks and its traffic workload. Because of the complexity of realistic quantitative analysis of network behavior, certain aspects of network performance studies can only be conducted effectively using simulation.

Optimization: Network performance optimization can be called corrective when a solution to a problem is made, or perfective, where an improvement to the network performance is made, even when there is no problem. Many actions could be taken such as adding additional links, increasing link capacity or adding additional hardware. Planning for future improvement in the network (e.g. network design, network capacity or network architecture) is considered as a part of network optimization.

9. Criteria for selecting the best traffic route

Traditionally, there have been three parameters that describe the quality of a connection: bandwidth, delay, and packet loss. A connection with high bandwidth, low delay, and low packet loss is considered to be better than one with low bandwidth, high delay, and high packet loss. The following parameters can be considered when selecting the best traffic route:

Congestion: Congestion decreases the available bandwidth and increases delay and packet loss. It is important to avoid routes over congested paths.

Distance: Two routes may have different paths. Some networks interconnect only at relatively few locations, so they may have to transport traffic over long distances to get it to its destination. Others have better interconnection, so the traffic does not have to take a detour. There may be reasons not to prefer the more direct route, such as lower bandwidth or congestion, but generally a shorter geographic path is better.

Hops: The number of hops (e.g. routers) that shows up on the path to the destination increases the delay. Each hop potentially adds additional delay, because packets have to wait in a queue before they are transmitted, and the extra equipment in a path means that a failure somewhere along the way is more likely. So, paths with fewer hops are better.

10. Congestion control

Congestion in a packet switching network is a state in which the performance of the network degrades because of the saturation of network resources. Congestion could result in degradation of service quality to users. To avoid congestion, certain mechanisms have to be provided; such mechanisms are usually called congestion control.

10.1 Categories of congestion control

Congestion control policies can be categorized differently based on the objective of the policy, the time period of applying the policy, and the action taken to avoid congestion. In the following we will explain some of these policies.

10.1.1 Response time scale

Response time scale can be categorized as one of the following: long, medium and short. **In the long time scale**, expansion of the network capacity is considered. This expansion is based on estimates of future traffic demands, and traffic distribution. Because the network elements are expensive, upgrades take place in a long time scale between week to month or years. **In the medium time scale**, network control policies are considered (e.g. adjusting the routing protocol parameters to reroute traffic from a congested network node). These policies are mostly based on a measurement, and the actions are applied during a period of minutes to days. **In the short time scale**, packet level processing and buffer management functions in routers are considered (e.g. active queue control schemes in TCP traffic using Random Early Detection (RED)).

10.1.2 Reactive versus preventive

In reactive congestion control, congestion recovery takes place to restore the operation of a network to its normal state after congestion has occurred. Control policies react to existing congestion problems to remove or reduce them. In preventive congestion control, keeping the operation of a network at or near the point of maximum power is the main objective, so congestion will never occur. Control policies applied to prevent congestion are based on estimates and predictions of possible congestion appearance.

10.1.3 Supply side versus demand side

Increasing the capacity in the network is called a supply side congestion control. Supply side control is achieved by increasing the network capacity or balancing the traffic, (e.g. capacity planning to estimate traffic workload). For demand side control, policies are applied to regulate the offered traffic to avoid congestion (e. g. traffic shaping mechanism is used to regulate the offered load).

10.2 Control policies

Different congestion control policies have been proposed to deal with congestion in networks. Generally speaking, these policies differ in the use of control messages. The following will describe some of them.

Source Quench: Source Quench is the current method of congestion control in the Internet. When a network node responds to congestion by dropping packets, it could send an Internet Control Message Protocol Source Quench message (ICMP) to the source, informing it of packet drop. The drawback of this policy is that it is a family of varied policies. The major gateway manufacturers have implemented various source quench methods. This variation makes the end-system user, on receiving a Source Quench, uncertain of the cause in which the message was issued (e.g. heavy congestion, approaching congestion, burst causing massive overload).

Random Drop: Random Drop is a congestion control policy intended to give feedback to users whose traffic congests the gateway by dropping packets. In this policy, randomly selected packets for a particular user, from incoming traffic, will be dropped. A user generating much traffic will have much more packets drop than the user who generate little traffic. The selection of packets drop in this policy is completely uniform. Random Drop can be categorized as Congestion recovery or congestion avoidance.

Congestion recovery tries to restore an operating state, when demand has exceeded capacity. Congestion avoidance is preventive in nature. It tries to keep the demand on the network at or near the point of maximum power, so that the congestion never occurs.

Congestion Indication: The so-called Congestion Indication policy uses a similar technique as the Source Quench policy to inform the source gateway of congestion. The information is communicated in a single bit. The Congestion Experienced Bit (CEB) is set in the network header of the packets already being forwarded by a gateway. Based on the value of this bit, the end-system user should make an adjustment to the sending window. The Congestion Indication policy works based upon the total demand on the gateway. For fairness the total number of users causing the congestion is not considered. Only users who are sending more than their fair share (allowed bandwidth) should be asked to reduce their load, while others could attempt to increase their load where possible.

Fair Queuing: Fair queuing is a congestion control policy where separate gateway output queues are maintained for individual end-systems on a source-destination-pair basis. When congestion occurs, packets are dropped from the longest queue. At the gateway, the processing and link resources are distributed to the end-systems on a round-robin basis. Round-robin is an arrangement of choosing all elements in a group equally in a circular. Equal allocations of resources are provided to each source-destination pair.

A Bit-Round Fair Queuing algorithm was an improvement over the fair queuing. It computes the order of service to packets using their lengths, by using a technique that emulates a bit-by-bit round-robin discipline. In this case, long packets do not get an advantage over short packets. Otherwise the round-robin would be unfair.

Stochastic Fairness Queuing (SFQ) is a similar mechanism to Fair Queuing. SFQ looks up the source-destination address pair in the incoming packets and locates the appropriate queue that packet will have to be placed in. It uses a simple hash function to map from the source-destination address pair to a fixed set of queues. The price paid to implement SFQ is that it requires a potentially large number of queues.

11. MPLS, and GMPLS traffic engineering

Network control (NC) can be classified as centralized or distributed. In centralized network control, the route control and route computation commands are implemented and issued from one place. Each node in the network communicates with a central controller and it is the controller's responsibility to perform routing and signaling on behalf of all other nodes. In a distributed network control, each node maintains partial or full information about the network state and existing connections. Each node is responsible to perform routing and signaling. Therefore, coordination between nodes is required to alleviate the problem of contention.

Since its birth, the Internet (IP network) has employed a distributed NC paradigm. The Internet NC consists of many protocols. The functionality of resource discovery and management, topology discovery, and path computation and selection are the responsibility of routing protocols. Multiprotocol label switching (MPLS) has been proposed by IETF, to enhance classic IP with virtual circuit-switching technology in the form of label switched path (LSP). MPLS is well known for its TE capability and its flexible control plane.

Then, IETF proposed an extension to the MPLS-TE control plane to support the optical layer in optical networks; this extension is called the Multiprotocol Lambda Switching (MPλS) control plane. Another extension to MPLS was proposed to support various types of switching technologies. This extension is called Generalized Multi- Protocol Label Switching (GMPLS). GMPLS has been proposed in the Control and Measurement Plane working group in the IETF as a way to extend MPLS to incorporate circuit switching in the time, frequency and space domains.

11.1 MPLS traffic engineering architecture

Multiprotocol label switching (MPLS) is a hybrid technology that provides very fast forwarding at the cores and conventional routing at the edges. MPLS working mechanism is based on assigning labels to packets based on forwarding equivalent classes

(FEC) as they enter the network. A FEC identifies of a group of packets that share the same requirements for their transport. All packets in such a group are provided the same treatment en route to the destination.

Packets that belong to the same FEC at a given node follow the same path and the same forwarding decision is applied to all packets. Then packets are switched through the MPLS domain using simple label lookup. Each FEC may be given a different type of service. At each hop, the routers and switches use the packet labels to index the forwarding table to determine the next-hop router and a new value for the label. This new label replaces the old one and the packet is forwarded to the next hop. As each packet exits the MPLS domain, the label is stripped off at the egress router, and then the packet is routed using conventional IP routing mechanisms.

The router that uses MPLS is called a label switching router (LSR). A LSR is a high speed router that participates in establishment of LSPs using an appropriate label signaling protocol and high-speed switching of the data traffic based on the established paths.

MPLS-TE has the following components and functionalities, as shown in Figure 2:

1. The routing protocol (e.g. OSPF-TE, IS-IS TE), collects information about the network connectivity (this information is used by each network node to know the whole topology of the network) and carries resource and policy information of the network. The collected information is used to maintain: The so-called Link-state database which provides a topological view of the whole network and the TE database which stores resource and link utilization information. The databases are used by the path control component. A constrained Shortest Path First (CSPF) is used to compute the best path.
2. A signaling protocol (e.g. RSVP-TE or CR-LDP) is used to set up LSP along the selected path through the network. During LSP setup, each node has to check whether the requested bandwidth is available. This is the responsibility of the link admission control that acts as an interface between the routing and signaling protocol. If bandwidth is available, it is allocated. If not, an active LSP might be preempted or the LSP setup fails.

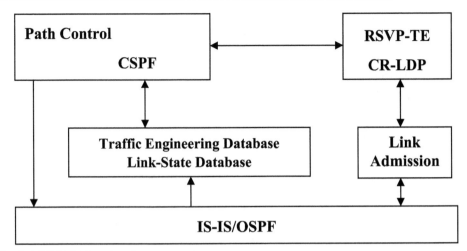

Fig. 3. MPLS-TE functional components.

11.2 MPλS/GMPLS control plane

IETF proposed an extension to the MPLS-TE control plane to support optical layers in optical networks; this extension is called the multiprotocol lambda switching (MPλS) control plane. In an MPLS network, the label-switching router (LSR) uses the label swapping paradigm to transfer a labeled packet from an input port to an output port. In the optical network, the OXC uses switch matrix to switch the data stream (associated with the light path) from an input port to an output port. In both LSR and OXC, a control plane is needed to discover, distribute, and maintain state information and to instantiate and maintain the connections under various TE roles and policies.

The functional building blocks of the MPλS control plane are similar to the standard MPLS-TE control plane. The routing protocol (e.g. OSPF or IS-IS) with optical extensions, is responsible for distributing information about optical network topology, resource availability, and network status. This information is then stored in the TE database. A constrained-based routing function acting as a path selector is used to compute routes for LSPs through mesh network. Signaling protocols (e.g. RSVP-TE or CR-LDP) are then used to set up and maintain the LSPs by consulting the path selector.

Another extension to the MPLS control plane is proposed to support various types of optical and other switching technologies. This extension is called Generalized Multi-Protocol Label Switching (GMPLS). In the GMPLS architecture, labels in the forwarding plane of Label Switched Routers (LSRs) can route the packet headers, cell boundaries, time slots, wavelengths or physical ports. The following switching technologies are being considered, as shown in Figure 3.

Packet switching: The forwarding mechanism is based on packet. The networking gear is an IP router.

Layer 2 switching: The forwarding mechanism is based on cell or frame (Ethernet, ATM, and Frame Relay).

Time-division multiplexing (time slot switching): The forwarding mechanism is based on the time frames with several slots and data is encapsulated into the time slots (e.g. SONET/SDH).

Lambda switching: λ switching is performed by OXCs.

Fiber switching: Here the switching granularity is a fiber. The networkings gears are fiber switch capable OXCs.

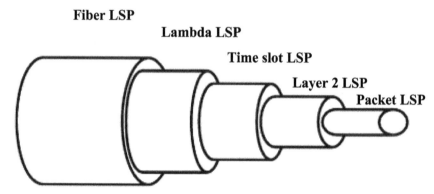

Fig. 4. GMPLS Label – Stacking Hierarchy.

The difference between MPλS and GMPLS is that the MPλS control plane focuses on Lambda switching, while GMPLS includes almost the full range of networking technologies.

12. MPLS traffic engineering features

1. **Explicit routes:** MPLS supports setting up of explicit routes, which can be an important tool for load balancing and satisfying other objectives so as to steer traffic away from particular paths. It is a very powerful technique which potentially can be useful for a variety of purposes. With pure datagram routing the overhead of carrying a complete explicit route with each packet is prohibitive. However, MPLS allows the explicit route to be carried only at the time that the label switched path is set up, and not with each packet. This implies that MPLS makes explicit routing practical. This in turn implies that MPLS can make possible a number of advanced routing features which depend upon explicit routing.
    An explicitly routed LSP is an LSP where, at a given LSR, the LSP next hop is not chosen by each local node, but rather is chosen by a single node (usually the ingress or egress node of the LSP). The sequence of LSRs followed by an explicit routing LSP may be chosen by configuration, or by an algorithm performed by a single node (for example, the egress node may make use of the topological information learned from a link state database in order to compute the entire path for the tree ending at that egress node).
    With MPLS the explicit route needs to be specified at the time that Labels are assigned, but the explicit route does not have to be specified with each L3 packet. This implies that explicit routing with MPLS is relatively efficient (when compared with the efficiency of explicit routing for pure datagram).

Explicit routing may be useful for a number of purposes such as allowing policy routing and/or facilitating traffic engineering.

2. **Path preemption:** Some tunnels are more important than others. Say for example, a VoIP tunnel and data tunnel may compete for same resources, in which case VoIP tunnel is given a higher priority and data tunnel is made to recalculate a path or just drop, if no path is available. Tunnels have 2 priorities: setup priority and hold priority. Each can have a value from 0 to 7 and the higher the priority numbers the lower the tunnels importance. The setup priority is used when setting up a tunnel and is compared with the hold priority of already established ones. If the setup priority is higher than the hold priority of established tunnel, then established tunnel is preempted.

3. **Fast Re-route:** In case of a link failure, interior gateway protocols may take of the order of 10 seconds to converge. Fast reroute involves pre-signaling of backup path along with the primary path. The protection may be path protection (end-to-end) or local protection which may be further differentiated into link protection and node protection.

Fast reroute is a Multiprotocol Label Switching (MPLS) resiliency technology to provide fast traffic recovery upon link or router failures for mission critical services. Upon any single link or node failures, it could be able to recover impacted traffic flows in the level of 50 ms.

Backup path can be configured for:

1. **Link protection:** a link protection model each link (or subset links) used by an LSP is provided protection by pre-established backup paths.

2. **Node protection:** In a node protection model each node (or subset of nodes) used by an LSP is provided protection by pre-established backup paths.

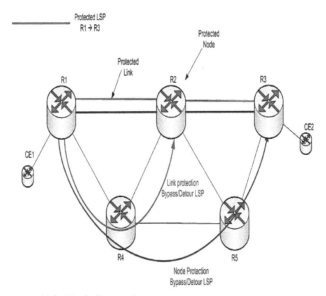

Fig. 5. Link Protection V/s Node Protection

13. References

[1] J. Malcolm, J. Agogbua, M. O'Dell and J. McManus, "Requirements for Traffic Engineering Over MPLS," IETF RFC 2702, September 2004.

[2] T. Li and Y. Rekhter, "A Provider Architecture for Differentiated Services and Traffic Engineering (PASTE)," IETF RFC 2430, October 2006.

[3] M. Allman, V. Paxon, and W. Stevens. TCP congestion control. Request for Comments (Standards Track) RFC 2581 April 1999. URL:http://www.ietf.org/rfc/rfc2581.txt.

[4] D. Awduche, J. Malcolm, J. Agogbua, J. McManus "Requirements for Traffic Engineering over MPLS (RFC 2702)" http://rfc-2702.rfc-list.net/rfc-2702.htm Sept 1999.

[5] V. Alwayn, "Advanced MPLS Design and Implementation"ISBN 1-58705-020-X.

[6] L. Andersson, P. Doolan, N. Feldman, A. Fredette, B. Thomas "LDP Specification (RFC 3036)" http://rfc-3036.rfc-list.net/ January 2001.

[7] Li, T. and Y. Rekhter, "Provider Architecture for Differentiated Services and Traffic Engineering (PASTE)", RFC 2430, October 1998.

Permissions

The contributors of this book come from diverse backgrounds, making this book a truly international effort. This book will bring forth new frontiers with its revolutionizing research information and detailed analysis of the nascent developments around the world.

We would like to thank Narottam Das, for lending his expertise to make the book truly unique. He has played a crucial role in the development of this book. Without his invaluable contribution this book wouldn't have been possible. He has made vital efforts to compile up to date information on the varied aspects of this subject to make this book a valuable addition to the collection of many professionals and students.

This book was conceptualized with the vision of imparting up-to-date information and advanced data in this field. To ensure the same, a matchless editorial board was set up. Every individual on the board went through rigorous rounds of assessment to prove their worth. After which they invested a large part of their time researching and compiling the most relevant data for our readers. Conferences and sessions were held from time to time between the editorial board and the contributing authors to present the data in the most comprehensible form. The editorial team has worked tirelessly to provide valuable and valid information to help people across the globe.

Every chapter published in this book has been scrutinized by our experts. Their significance has been extensively debated. The topics covered herein carry significant findings which will fuel the growth of the discipline. They may even be implemented as practical applications or may be referred to as a beginning point for another development. Chapters in this book were first published by InTech; hereby published with permission under the Creative Commons Attribution License or equivalent.

The editorial board has been involved in producing this book since its inception. They have spent rigorous hours researching and exploring the diverse topics which have resulted in the successful publishing of this book. They have passed on their knowledge of decades through this book. To expedite this challenging task, the publisher supported the team at every step. A small team of assistant editors was also appointed to further simplify the editing procedure and attain best results for the readers.

Our editorial team has been hand-picked from every corner of the world. Their multi-ethnicity adds dynamic inputs to the discussions which result in innovative outcomes. These outcomes are then further discussed with the researchers and contributors who give their valuable feedback and opinion regarding the same. The feedback is then collaborated with the researches and they are edited in a comprehensive manner to aid the understanding of the subject.

Apart from the editorial board, the designing team has also invested a significant amount of their time in understanding the subject and creating the most relevant covers. They scrutinized every image to scout for the most suitable representation of the subject and create an appropriate cover for the book.

The publishing team has been involved in this book since its early stages. They were actively engaged in every process, be it collecting the data, connecting with the contributors or procuring relevant information. The team has been an ardent support to the editorial, designing and production team. Their endless efforts to recruit the best for this project, has resulted in the accomplishment of this book. They are a veteran in the field of academics and their pool of knowledge is as vast as their experience in printing. Their expertise and guidance has proved useful at every step. Their uncompromising quality standards have made this book an exceptional effort. Their encouragement from time to time has been an inspiration for everyone.

The publisher and the editorial board hope that this book will prove to be a valuable piece of knowledge for researchers, students, practitioners and scholars across the globe.

List of Contributors

Ricardo Barrios and Federico Dios
Department of Signal Theory and Communications, Technical University of Catalonia, Spain

Jian Zhao and Andrew D. Ellis
Photonic Systems Group, Tyndall National Institute & Department of Physics, University College Cork, Ireland

Abdulsalam Alkholidi and Khalil Altowij
Faculty of Engineering, Electrical Engineering Department, Sana'a University, Sana'a, Yemen

Inderpreet Kaur
Rayat and Bahra Institute of Engineering, Mohali, India

Neena Gupta
PEC University of Technology (Formally Punjab Engineering College), Chandigarh, India

Takashi Hiraga and Ichiro Ueno
National Institute of Advanced Industrial Science & Technology, Japan

Marija Furdek and Nina Skorin-Kapov
University of Zagreb, Croatia

Narottam Das
Department of Electrical and Computer Engineering, Curtin University, Australia
School of Engineering, Edith Cowan University, Australia

Hitoshi Kawaguchi
Graduate School of Materials Science, Nara Institute of Science and Technology, Japan

Jurgis Porins, Vjaceslavs Bobrovs and Girts Ivanovs
Riga Technical University, Institute of Telecommunications, Latvia

Kiazand Fasihi
Golestan University, Iran

Mahesh Kumar Porwal
Shrinathji Institute of Technology & Engineering, Nathdwara (Rajasthan), India

Printed in the USA
CPSIA information can be obtained
at www.ICGtesting.com
JSHW011815301024
72690JS00002B/88